NEHA

National Environmental Health Association

PROFESSIONAL FOOD MANAGER 3RD EDITION

Updated to the 2011 supplement
of the 2009 FDA Food Code

NEHA
EDUCATION & TRAINING

WILEY

John Wiley & Sons, Inc.

This book was set in 9/12 Helvetica Neue . Page layout by Maureen Eide and printed and bound by Courier. The cover was printed by Courier.

This book is printed on acid-free paper. ∞

Published by John Wiley & Sons, Inc., Hoboken, New Jersey.

Published simultaneously in Canada.

Evaluation copies are provided to qualified academics and professionals for review purposes only, for use in their courses during the next academic year. These copies are licensed and may not be sold or transferred to a third party. Upon completion of the review period, please return the evaluation copy to Wiley. Return instructions and a free of charge shipping label are available at www.wiley.com/go/returnlabel. Outside of the United States, please contact your local representative.

For general information on our other products and services, or technical support, please contact our Customer Care Department within the United States at 800-762-2974, outside the United States at 317-572-3993 or fax 317-572-4002.

Wiley also publishes its books in a variety of electronic formats. Some content that appears in print may not be available in electronic books. For more information about Wiley products, visit our Website at www.wiley.com.

Library of Congress Cataloging in Publication Data

ISBN: 978-1-118-38087-1

Printed in the United States of America

10 9 8 7 6 5 4 3 2

CONTENTS

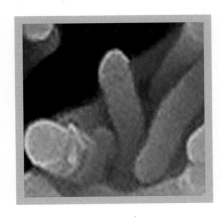

III

PREFACE

Food handling and protection is a critical part of the safety and success of a food service operation. An effective food safety program prevents foodborne illnesses and disease outbreaks, keeping guests as well as employees safe. Knowing food safety risks and how to prevent them is the first step toward creating a healthy environment. For food managers, setting high safety standards and adapting as industry guidelines evolve are part of each workday. Whether you are currently seeking food manager certification or are already a certified professional, knowing and applying best practices in food safety helps you to control risk and focus on prevention responsibly.

The National Environmental Health Association (NEHA)—the leader in environmental health and protection training—has created **Professional Food Manager, Third Edition** to provide essential food safety information for managers and soon-to-be managers looking to become certified, as well as professionals looking to refresh their knowledge of current food safety issues. This book presents a wealth of information, including the most up-to-date food safety trends, key principles of food safety management, health department guidelines, temperature recommendations and basic sanitation procedures to ensure healthy food handling practices.

What's New for the Third Edition

The **Third Edition** continues to offer streamlined, easy-to-understand information on creating and maintaining a safe food service operation. It includes updates and enhancements, such as statistics on foodborne illness, managing suppliers, monitoring food temperatures and purchasing equipment approved by NSF International and Underwriters Laboratories (UL) for industry safety.

In this latest edition, you'll also find:

- Current information, updated to the 2011 Supplement of the 2009 Food and Drug Administration Food Code
- Information on identifying and preventing contaminants from entering your work environment
- Additional tips on the handling of raw meat, what to expect from your food supplier and much more
- A new, visually appealing design

Supplementary Materials

In addition to the print book, qualified adopters of the NEHA materials have access to a wide array of resource materials to support their classes through an easy-to-navigate companion website at www.wiley.com/college/neha, including:

- **Instructor's Manual/Activity Guide,** which contains chapter quizzes, true/false questions, matching exercises, fill-in-the-blank questions and more
- **PowerPoint® Presentation** with Discussion Notes
- **Course Pre-Test** and Answer Key
- **Practice Examination** and Answer Key
- **Class Syllabi** for 8-hour and 16-hour classes
- **Case Studies** with Discussion Questions
- **Test Bank,** specifically formatted for Respondus, an easy-to-use software program for creating and managing exams that can be easily printed to paper or published directly to Blackboard, WebCT, Desire2Learn, eCollege, ANGEL and other eLearning systems.

About the National Environmental Health Association

Headquartered in Denver, Colorado, the National Environmental Health Association (NEHA) is an educational and professional organization that represents and supports professionals in the field of environmental health and protection throughout North America. Founded in 1937, and now with 53 affiliated associations worldwide, NEHA is one of the most highly regarded associations representing environmental health professionals.

NEHA's mission, "to advance the environmental health and protection professional for the purpose of providing a healthful environment for all" is as relevant today as it was when the organization was founded. The basis for the association's activities is the belief that the professional who is educated and motivated is the professional who will make the greatest contribution to healthful environmental goals. Accordingly, great emphasis is placed on providing learning opportunities. One of NEHA's chief educational programs is Food Safety Training.

NEHA works closely with health departments and keeps abreast of food safty trends, so you receive the most up-to-date, effective training on the market. By working with NEHA materials, you will be able to develop long-standing relationships with your regulatory officials. Moreover, these materials are presented in ways that ensure both managers and handlers will remember the food safety principles long after training has taken place.

EDUCATION & TRAINING

NEHA

National Environmental Health Association

PROFESSIONAL FOOD MANAGER

MANAGING FOOD SAFETY AND TRAINING

INTRODUCTION TO FOOD SAFETY

LESSON 1

After completing this lesson, you should be able to:

- Define safe food
- Differentiate between a foodborne illness and a foodborne disease outbreak
- Describe causes of contamination and prevention methods
- Describe the benefits of good food safety standards

> **TERMS TO KNOW**

ACUTE ILLNESS An illness that develops rapidly and produces symptoms quickly after infection.

FOOD SAFETY The measures and conditions necessary to control hazards and to ensure fitness for human consumption of a foodstuff, taking into account its intended use.

FOODBORNE DISEASE OUTBREAK An occurrence of two or more cases of the same illness resulting from the consumption of the same food.

FOODBORNE ILLNESS An acute illness resulting from eating contaminated food, with symptoms including abdominal pain, diarrhea, vomiting and nausea.

SAFE FOOD Food that is free of contaminants.

What Is Safe Food?

Safe food is food that is free of contaminants. **Food safety** involves all the measures necessary to ensure the safety and wholesomeness of food at all stages, from receipt of raw materials to sale to the customer. The most important food safety principles are:

- Rejecting contaminated food or food from suspect sources
- Protecting food from contamination
- Preventing multiplication of microorganisms
- Destroying microorganisms
- Discarding or removing unsafe or contaminated food

Common symptoms of a foodborne illness are abdominal pain, diarrhea, vomiting and nausea.

Foodborne Illness

The primary goal of a food safety program is to prevent foodborne illnesses and foodborne disease outbreaks. A **foodborne illness** is an **acute illness** resulting from eating contaminated food, with symptoms including abdominal pain, diarrhea, vomiting and nausea. A **foodborne disease outbreak** is the occurrence of two or more cases of the same illness resulting from the consumption of the same food.

Foodborne illness in the United States is a major cause of personal distress, preventable illness and death, and avoidable economic burden. Scallan et al. (2011a,b) estimated that foodborne diseases cause approximately 48 million illnesses, 128,000 hospitalizations, and 3,000 deaths in the United States each year. The occurrence of approximately 1,000 reported disease outbreaks (local, regional, and national) each year highlights the challenges of preventing these infections.

Most foodborne illnesses occur in persons who are not part of recognized outbreaks. For many victims, foodborne illness results only in discomfort or lost time from the job. For some, especially preschool age children, older adults in health care facilities, and those with impaired immune systems, foodborne illness is more serious and may be life threatening.

Causes of Foodborne Disease Outbreaks

Most foodborne disease outbreaks are caused by a failure to implement good food safety principles. This failure may result from ineffective supervision, lack of managerial or staff commitment or insufficient training, leading to the negligence or ignorance of staff involved in the preparation, storage, distribution or processing of food.

Preventing Foodborne Disease Outbreaks

Understanding the causes of foodborne disease outbreaks as a manager will help you prevent them. Prevention of most foodborne disease outbreaks is relatively simple.

- Cook food thoroughly
- Protect food from contamination
- Keep hot foods hot
- Keep cold foods cold

Cook food thoroughly

Protect food

Keep hot foods hot

Keep cold foods cold

Benefits of Food Safety

Benefits of Good Food Safety	Consequences of Poor Food Safety
Satisfied customers	Foodborne disease outbreaks
Good reputation	Food contamination
Increased business	Customer complaints
Legal compliance	Pest infestations
Minimal food waste	Food waste
Good working conditions	Closure of premises
Higher staff morale	Fines and civil action
Reduced staff turnover	Loss of business
Increased productivity	Lower profits
Better relationship with enforcement officers	
Higher profits	

If you cook food thoroughly, protect it from contamination, keep it hot or cool it rapidly and then keep it cold, you will prevent most foodborne disease outbreaks. To ensure your business is continually working toward outbreak prevention, you will need to set up and oversee a number of systems, procedures and checklists and coach and train your staff.

Food Safety in the Media

As times change, so do your responsibilities as a manager of a food establishment. With the increase in mass media outlets and electronic media, often times, foodborne disease outbreaks receive a high level of media attention. Therefore, it's vital not only that you provide safe food in a safe environment but also that you are prepared to deal with client questions and concerns regarding the safety of the food you provide.

MANAGEMENT

LESSON 2

After completing this lesson, you should be able to:

- Apply basic management principles to food safety management
- Create a food safety culture in your establishment
- Outline standards for a safety policy for your workplace

>TERMS TO KNOW

EXCLUSION Requiring a worker to leave the food establishment as a result of specific illnesses, symptoms or exposure to certain diseases.

FOOD SAFETY MANAGEMENT SYSTEM The policies, procedures, practices, controls and documentation that ensure that food sold by a food business is safe to eat and free from contaminants.

FOOD SAFETY POLICY A company's commitment to producing safe food, providing satisfactory premises and equipment and ensuring that legal responsibilities are met and appropriate records maintained. The document outlines management responsibilities and is used to communicate standards to staff.

PERSON IN CHARGE (PIC) An individual responsible for a food establishment, or the department or area of a food establishment.

The new update to the FDA *Food Code* requires that at least one food establishment employee with management and supervisory responsibility be a Certified Food Protection Manager (CFPM). That means it is very important for you to know and understand all of the information in this manual. After the course, please keep this book with you for reference.

Management Overview

Let's start with a quick re-cap on basic management principles. Management is the process of getting activities completed efficiently and effectively with and through other people. As a manager, it is your job to monitor employee actions and ensure that your employees are not only trained correctly in food safety principles, but that they are following through on correct methods and procedures. There are four basic tiers of management: plan, organize, direct and monitor.

Plan

Management always starts with planning. You need to understand where you are currently, understand where you need to be (that is, the required standards) and figure out the best way to get there.

Organize

Once you have a plan, you will need to make it happen. Ask yourself:

- Are all the pieces of the plan in place?
- Has my staff received sufficient training?
- Is my staff motivated?
- Does my staff have the proper equipment?

Direct

Think of directing as conducting an orchestra. You've given your staff the plan, and you've organized your way to plan implementation. Now it's time to flip the "ON" switch.

Monitor

Once the plan is in motion, you have to keep an eye on things. During the monitoring process, you need to make any necessary adjustments, coach your staff and provide your staff with feedback.

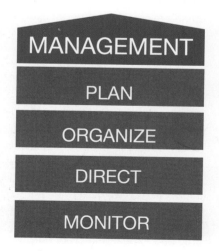

Food Safety Management

How do we apply the principles of planning, organizing, directing and monitoring to **food safety management system**? First set the standards you want to achieve, and then provide resources and implement systems that will help you achieve those standards. The standards and systems need to be clearly communicated to all staff. Your team will require motivation and training. This includes giving them specific instructions and on-the-job training to develop competency. Your team will also require solid supervision, especially at first, to make sure that the training is effective. You complete your circle of responsibility by continually monitoring, analyzing, adjusting and improving performance.

CIRCLE OF RESPONSIBILITY

1. Set the standard
2. Provide resource and implement systems
3. Communicate to all staff
4. Motivate and train
5. Supervise
6. Monitor, analyze, adjust, and improve performance continually

Creating a Food Safety Culture

Good management will lead to a motivated staff, a great working environment and the start of a food safety culture. In addition to good management, it is also very important to:

- Lead by example. For instance, you should always wear the appropriate protective clothing, wash your hands upon entering a food preparation or serving room and follow correct procedures, just as you expect your staff to follow them.
- Keep up-to-date with food safety matters, including new regulations and company procedures.
- Communicate effectively, both with your staff and with the business owners or other company representatives.
- Build other skills. Beyond the principles outlined here, there are other skills that will help you in your job, such as leadership, recruiting, coaching, team building and delegation. You can build these skills by taking one of the many courses available on these topics.

Standards are needed to ensure consistency and provide a reference point.

The food safety policy outlines management's responsibilities and is also used to communicate standards to the staff.

Standards

Now that you understand general management principles and how they relate to food safety, let's take a look at how to put this knowledge to work through standards and policies.

Standards are needed to ensure consistency and provide a reference point to determine when a target has been achieved. Standards are also needed to determine that a task, such as cleaning a meat slicer, has been completed satisfactorily. Standards can relate to premises, people or products and may be used in several different ways. There are a variety of standards, including voluntary standards such as in-house standards, and legal standards such as federal, state and local regulations. Standards can be set by a variety of individuals and organizations, including you, your company, your customer, the government, trade unions, independent standards authorities, professional associations, local authorities and food safety authorities.

Policies

A company's standards should be clearly set out in its food safety policy. The **food safety policy** outlines management's responsibilities and is also used to communicate standards to the staff. The policy is effectively a commitment to produce safe food, provide satisfactory premises and equipment and ensure that legal responsibilities are met and appropriate records maintained. A food safety policy should include procedures for:

- Approval and monitoring of suppliers and delivery of products
- Personal hygiene, health screening and reporting of illnesses
- Informing and supervising staff
- Implementing an effective training program
- Effective temperature control and monitoring practices
- Integrated pest management
- Cleaning and sanitizing, including cleaning schedules
- Waste management
- Customer complaints
- Foodborne disease outbreaks
- Information on food allergies
- Quality control and assurance
- Handling inspections and inspectors

Person in Charge

The FDA *Food Code* defines the **person in charge (PIC)** as: "the individual present at a food establishment who is responsible for the operation at the time of inspection." In many cases, you will be the person in charge. However, when you are not present at your food establishment, it's important that you have identified a person in charge to oversee safe food service and deal with questions or concerns regarding food safety. In addition to ensuring your establishment is serving safe food, the person in charge must demonstrate a wide range of knowledge in day-to-day operations and during an inspection.

DELIVERING TRAINING

LESSON 3

After completing this lesson, you should be able to:

- List the stages of training
- Explain different delivery methods
- Describe the elements of a complete training program

Stages of Training

Training comprises four main stages:

- Motivate your staff to learn
- Provide knowledge and pass on practical skills via demonstrations
- Supervise the staff while they are practicing the tasks they've learned
- Test their competency by checking that the task has been performed satisfactorily

There are many different training methods you can use to convey knowledge and skills. It is beneficial to use a mixture of methods to ensure that the training is effective.

Consider employees' learning styles when establishing training methods.

Delivery Methods

When determining what delivery methods will work for you, recognize that training must be delivered in a manner that is appropriate to the training content, the workplace and the audience.

For example, it would not be appropriate to give examples throughout the training that had no bearing to the staff member's workplace, or to give in-depth microbiological explanations to front-line staff members.

When planning training and methods of delivery, you should also consider:

- The employee's learning style. For example, is the person very hands-on, do they like to go through written documents on their own or do they find working at their own pace on the computer less threatening?
- Whether there are any barriers to learning. This would include such things as language, attitude, the use of jargon or whether the learner has any learning difficulties or disabilities.

Share benefits of training

Employees need to see the value in training. Make it clear how the training will make their jobs easier and help them work faster.

Offer hands-on activities

Devote about two-thirds of training time to hands-on activities that allow staff members to practice their new skills.

Give feedback

This gives you the opportunity to provide feedback. Remember to keep feedback immediate, specific, and positive, and offer it for both correct and incorrect behaviors.

Keep sessions short

Break your training down into small, manageable chunks of information. Short sessions of no more than 45 minutes are always best to ensure retention.

Elements of a Complete Training Program

To ensure that you apply the four main stages of training to your food safety training program, each of the following elements must be present:

- You must have clearly defined objectives that can be measured.
- You must ensure that the training you're offering supports the identified objectives.
- You must evaluate your employees after completion of the training to make sure the objectives have been met.
- You must foster a work environment that allows and encourages employees to put their training into practice.
- **You must lead by example.** This reinforces that your company's commitment to food safety is not just talk; management's actions back it up.

Training Records

Your training program isn't complete unless you document it. Therefore, you should keep records of all food safety training that you conduct. This includes training for new employees, as well as any refresher training. You should keep a record of training for each individual employee. Not only are such records beneficial for reviewing and evaluating an employee's job performance, but it's also good to have this documentation should you need it for legal purposes.

New Hire Training

When hiring new staff members, be aware that they will not be familiar with your company's food safety procedures without the proper training. New employees need to understand the importance of following food safety principles right from the start. They should receive training on all aspects of food safety, including proper personal hygiene, safe food preparation, proper cleaning and sanitizing and pest recognition and prevention.

Proper personal hygiene

Safe food preparation

Proper cleaning and sanitizing

Pest recognition and prevention

Refresher Training

Refresher training is needed to ensure that all (not just new) employees stay up-to-date on all aspects of food safety. Refresher training is carried out as needed and whenever required, for example, following the implementation of a new system or equipment at the premises or the introduction of new legislation.

Training should be done at regular intervals, as dictated by the company's policies. It can also be dictated by circumstances such as an incident or complaint, your daily checks or training reviews. Use refresher training to ensure your staff's competency and to ensure that your staff's knowledge and skills are kept up to date.

ASSESSMENT

1. Which of the following is the term for an occurrence of two or more cases of the same symptoms after consumption of the same food?

   a. Foodborne illness
   b. Foodborne disease outbreak
   c. Foodborne complaint

2. Choose two. Which of the following are benefits of good food safety standards?

   a. Reduced costs from food waste
   b. Reduced costs from training food handlers
   c. Reduced costs from fines
   d. Reduced costs of purchasing cleaning chemicals

3. Which of the following has recently become another reason to practice good food safety principles?

   a. The discovery of new foodborne illnesses
   b. The fact that foodborne disease outbreaks have become illegal
   c. The increase in media attention and coverage

4. What should you do for your staff to reinforce the importance of high food safety standards?

   a. Motivate them
   b. Offer resources and training
   c. Provide solid supervision
   d. All of the above

5. Which of the following best describes a food safety policy?

   a. It should outline management's responsibilities and should also be used to communicate standards to the staff.
   b. It should contain the work hours policy in place at the establishment.
   c. It should be the only training provided to new employees.

6. Which three of the following should be included in a food safety policy?

   a. Personal-hygiene, health-screening and illness-reporting procedures
   b. Effective temperature control and monitoring practices
   c. Procedures for requesting time off from work
   d. The allowances made by the company's health insurance provider
   e. Procedures for handling foodborne disease outbreaks

7. Which of the following best describes the concept of management?

   a. Knowing the causes of foodborne disease outbreaks
   b. Making procedures
   c. Writing reports and hiring workers
   d. Getting activities completed efficiently and effectively with and through other people

8. True or False

   People, products and premises are all things that can be governed by food safety standards.

9. What is the main objective of food safety training on food premises?

   a. To make sure the premises are ready for a surprise visit from the health inspector
   b. To teach employees the importance of hand washing
   c. To improve working practices by changing staff behavior and attitude
   d. To keep the premises clean and pest-free

10. What are the four stages of training?

    a. Research, implement, deliver, review
    b. Motivate, teach, supervise, test
    c. Initiate, demonstrate, review, test
    d. Introduce, convey, supervise, correct

BIOLOGICAL CONTAMINATION

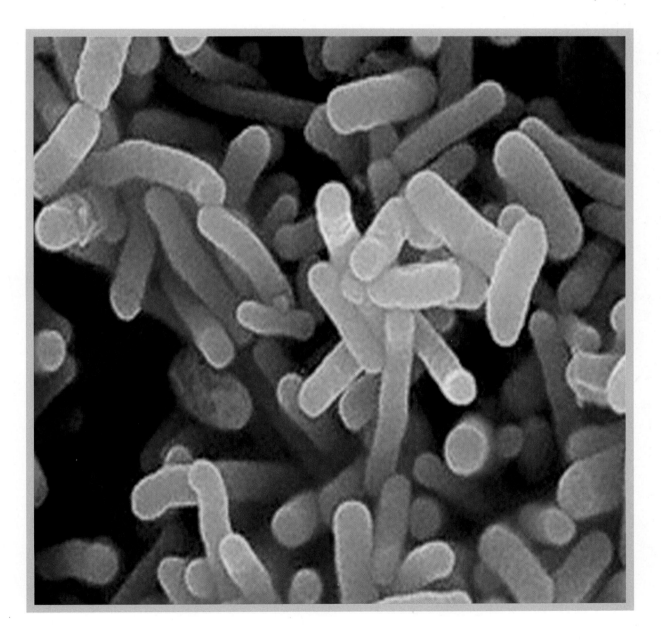

CONTAMINATION

LESSON 1

After completing this lesson, you should be able to:

- Define contamination
- Identify different biological contaminants
- Describe possible causes of biological contamination

What Is Contamination?

Contamination is the presence of physical, chemical or biological matter in or on food or the food environment. Food may be contaminated before delivery to a business, as a result of poor hygiene practices on the premises, or it may even be contaminated by customers, if not adequately protected while on display.

Biological Contaminants

Biological contaminants include bacteria, viruses, parasites and fungi. These microorganisms may be transferred to food from a variety of sources, such as people, raw food, contact surfaces, pests and refuse. **Biological contamination** usually occurs as a result of ignorance, inadequate space or poor structural design. It may also occur when food handlers take shortcuts or do not implement good hygiene practices. Biological contamination is the most common type of contamination and can have very serious consequences. In the early stages it will not be detectable, but it may result in food spoilage, food poisoning or even death.

BACTERIA

LESSON 2

After completing this lesson, you should be able to:

- Describe characteristics of bacteria
- Identify variables that affect bacterial growth
- Classify a bacterial illness

What Are Bacteria?

Bacteria are single-celled microscopic organisms and cannot be seen with the naked eye.

Large numbers of bacteria can be seen as small white or yellow spots, known as colonies, on growth media in laboratories.

Large numbers may also cause visible effects on food—for example, a slime layer on the surface of spoiled meat. Bacteria vary considerably in shape. Cocci are spherical, rods are sausage-shaped, spirochetes are spiral, and vibrios are comma-shaped. In simple terms, bacteria have a cell wall, which gives shape to the bacteria and holds together the cytoplasm and nuclear material inside. Flagella are attachments to the cell wall, which allow some bacteria to move in liquids. Some bacteria also have an outer slime layer called a capsule.

Bacterial multiplication

Bacteria multiply by binary fission, that is, they split into two. Each of these two daughter cells then grows to maturity and itself will divide again into two. Bacteria will continue to divide at regular intervals while conditions are suitable for their growth and multiplication. This state is described as the **vegetative state.**

The majority of bacteria are harmless and some are even essential to our health because they help us digest food and kill harmful microorganisms in our bodies. However, there are a small number of bacteria that spoil food, referred to as food spoilage bacteria, and others that cause illness, known as **pathogens**, which include food poisoning bacteria. Food poisoning bacteria are the most significant biological contaminant. Bacteria are found everywhere, including: on food, in water, in soil and air, on and in animals and pests and on and in people. Bacteria can be found on and in people, not necessarily because they are ill. People who show no symptoms of illness, but excrete food poisoning bacteria or simply carry them on their bodies, are referred to as **carriers** and are a source of food poisoning bacteria.

Spores

Some bacteria produce **spores**, which enable them to survive high temperatures. The phase when spores are formed is known as sporulation. Spores are activated by cooking and germinate during slow cooling. Spores are very difficult to kill and will also protect bacteria from other adverse conditions, such as drying and sanitization. Spores can survive for many years, without the need for food or water, but they do not multiply. It is only when favorable conditions return that they start to multiply again. *Clostridium perfringens, Clostridium botulinum* and *Bacillus cereus* all produce spores. A very small amount of *Clostridium botulinum* toxin can cause botulism, a deadly foodborne illness.

CHAPTER 2 | BIOLOGICAL CONTAMINATION

> ## TERMS TO KNOW

ACIDIC Having a pH level less than 7.0, as do foods such as vinegar and tomatoes.

AEROBE An organism that requires oxygen to multiply. Also referred to as an "aerobic organism."

ANAEROBE An organism that requires the absence of oxygen to multiply. Also referred to as an "anaerobic organism."

BACILLUS CEREUS An intoxication-causing bacteria commonly found in starchy foods and meat products. This type of bacteria produces two types of toxins: emetic (causes vomiting) and diarrheal (causes diarrhea). Each toxin causes a different type of illness.

BACTERIA Single-celled microorganisms with rigid cell walls that multiply by dividing into two (that is, by binary fission). Some bacteria cause illness and others cause food spoilage.

CARRIER A person who harbors, and may transmit, pathogenic organisms with or without showing signs of illness.

CLOSTRIDIUM BOTULINUM An intoxication-causing bacteria commonly found in soil, and therefore, in products that come from soil such as root vegetables. It is anaerobic, which means it grows without oxygen. Because there's no need for oxygen, *Clostridium botulinum* can also be found in improperly canned food.

CLOSTRIDIUM PERFRINGENS A bacteria that causes a mild infection from toxin-producing spores. It is anaerobic and can be found in soil, animal and human waste, dust, insects and raw meat.

DECLINE PHASE Bacterial growth phase in which the number of bacteria decreases because more bacteria are dying than are multiplying.

ENDOTOXIN Toxin (poison) present in the cell wall of many bacteria that is released on death of the bacteria.

EXOTOXIN Toxin (poison) usually produced during the multiplication of some bacteria. They are highly toxic proteins and are often produced in food.

FACULTATIVE ANAEROBE An organism that can multiply with or without the presence of oxygen.

CONTINUED

FAT TOM The acronym that lists the conditions that support the rapid growth of microorganisms. These conditions are food, acidity, time, temperature, oxygen and moisture.

GENERATION TIME The time between bacterial divisions.

INFECTION A disease caused by the release of endotoxins in the intestine of the affected person. Illnesses caused by infection normally have a longer onset time. It may take one or two days before the infection makes a person feel ill.

INTOXICATION An illness caused when bacteria produce exotoxins that are released into food. Illnesses caused by intoxication normally have a short onset time. Intoxication can also be caused by chemical residues and food additives.

LAG PHASE Bacterial growth phase when bacteria are not multiplying at all.

LOGARITHMIC PHASE Bacterial growth phase when bacteria multiply rapidly.

PATHOGEN A disease producing organism.

pH An index used as a measure of acidity/alkalinity, measured on a scale of 1 to 14. Acidic foods have pH values below 7 and alkaline foods above 7; a pH value of 7 is neutral.

STATIONARY PHASE Bacterial growth phase when the number of bacteria produced by multiplication equals the number of bacteria dying.

SPORE A resistant resting phase of bacteria, protecting them against adverse conditions such as high temperatures.

TEMPERATURE DANGER ZONE The temperature range at which most food-borne microorganisms rapidly grow. The temperature danger zone is 41°F to 135°F (5°C to 57°C).

VEGETATIVE STATE Capable of growing. Bacteria in the vegetative state continue to divide at regular intervals while conditions are suitable for growth and multiplication.

Bacteria Growth Variables: FAT TOM

There are six major variables that affect the growth of bacteria. These variables are food, acidity, time, temperature, oxygen and moisture. The first letter of each variable forms the acronym **FAT TOM**.

FOOD To multiply to dangerous levels, bacteria need the right sort of food, which will support bacterial multiplication. Bacteria get the nutrients they require from food that is high in protein, such as meat, fish and dairy products.

ACIDITY The amount of acidity in food will also affect the growth of bacteria. Food items that are **acidic** or high in sugar and salt don't usually support bacterial multiplication. **pH** is a unit of measurement that can be used to rate the acidity and alkalinity of food. The pH of a food item is measured on a scale of 1 to 14. Acidic foods have pH values below 7 and alkaline foods above 7; a pH value of 7 is neutral. Most bacteria prefer a pH between 4.6 and 7, that is, mildly acidic foods.

TIME Bacteria also need time at the right temperature in order to multiply to dangerous levels. The time between each bacterial division is known as the generation time. The average **generation time** of the common food poisoning bacteria, under optimum conditions, is around 20 minutes. However, some food poisoning bacteria can split into two every 10 minutes. This means that in just one hour and 40 minutes, 1,000 bacteria can become over one million.

TEMPERATURE Food poisoning bacteria generally multiply between 41°F and 135°F (5°C and 57°C)—the **temperature danger zone**. Most food poisoning bacteria multiply fastest between 70°F to 125°F (21°C to 52°C), especially at body temperature, which is around 98.6°F (37°C). Some bacteria, such as *Listeria*, multiply at temperatures below 41°F (5°C). None will multiply at temperatures above 135°F (57°C), but remember that spore-producing bacteria will not be completely destroyed at those higher temperatures.

OXYGEN Another variable affecting the growth of bacteria is oxygen. Oxygen is normally present in food, except food subjected to boiling or canning. Some bacteria, called **aerobes**, can multiply only in the presence of oxygen while others, called **anaerobes**, can multiply only when there is no oxygen. Other bacteria, however, can multiply with or without the presence of oxygen, and these are called **facultative anaerobes**. The process of cooking drives oxygen from the food. This may lead to ideal conditions for anaerobes, especially in the center of food such as stews.

MOISTURE The last major variable affecting the growth of bacteria is moisture. Moisture is wetness caused by a liquid, typically water. Bacteria require water to transport the nutrients from the food into their cell and take away waste products. Most food items contain sufficient moisture to enable bacteria to multiply.

The amount of moisture in a food item available to bacteria is usually measured in terms of water activity. The water activity of pure water is 1, and most bacteria multiply best in food with a water activity of between 0.95 and 0.99. For example, fresh meat has a water activity of between 0.95 and 1.

Food  Acidity  Time  Temperature  Oxygen  Moisture

Competition

There is an additional factor that plays an important role in bacterial multiplication: competition. If there are different bacteria present in food or the food environment, then they will be competing for the same available food, especially if their optimum conditions of temperature, pH and oxygen are similar. However, most food poisoning bacteria are not particularly competitive, which results in these bacteria gradually dying off unless they are present in high numbers.

Phases of bacterial growth

The phase in bacterial growth when bacteria multiply rapidly is called the **logarithmic phase**. However, bacteria are not in this phase all the time. If bacteria are not multiplying at all—for example, immediately following removal from a refrigerator —then they are said to be in the **lag phase**. If the number of bacteria produced by multiplication equals the number of bacteria dying, then they are in the **stationary phase**. And if the number of bacteria decreases as there are more bacteria dying than multiplying—for example, if the food is being cooked at the correct temperature or if there are few or no suitable nutrients—then the bacteria are said to be in their **decline phase**. So, depending on the availability of the right nutrients, moisture and temperature range, bacteria can have longer or shorter phases, or can remain in a phase for a long period of time.

Classifying Bacterial Illness

Bacterial food poisoning may be toxic or infectious. **Intoxication** is when bacteria produce and release exotoxins into food and cause illness. **Exotoxins** are poisons usually produced during the multiplication of some bacteria. They are highly toxic proteins and are often produced in food. Illnesses caused by intoxication normally have a short onset time. Intoxication can also be caused by chemical residues and food additives.

Infection is when bacteria release endotoxins in the intestine of the affected person and cause illness. An **endotoxin** is a poison present in the cell wall of many bacteria that is released on death of the bacteria. Illnesses caused by infection normally have a longer onset time. It may take one or two days before the infection makes a person feel ill.

Types of Bacteria

Bacteria	Sources	Common Food Vehicles	Symptoms	Onset Time	Duration	Specific Controls
Clostridium botulinum	• Soil • Sediment • Intestinal tracts of fish and mammals	• Sausages • Meat products • Canned vegetables • Seafood products	• Weakness, vertigo • Abdominal distention • Double vision • Difficulty speaking and swallowing	18 to 36 hours	Several months	• Pay close attention to time and temperature guidelines • Be careful not to use the contents of damaged cans
Staphylococcus aureus	• Human skin, nose and hands; boils and cuts • Dust • Raw milk from cows or goats	• Milk and dairy products • Deli salads • Desserts • Custards • Cooked meat and poultry	• Nausea • Vomiting • Abdominal cramping • Diarrhea	30 minutes	2 days	• Use proper hand-washing techniques • Cover cuts on arms and hands • Pay attention to time and temperature guidelines
Bacillus cereus	• Starchy foods (emetic toxin) • Meat products (diarrheal toxin)	• Rice • Corn • Potatoes • Meat products	EMETIC • Nausea • Vomiting DIARRHEAL: • Abdominal cramps • Watery diarrhea	EMETIC: .5 to 6 hours DIARRHEAL: 6 to 15 hours	EMETIC: Less than 24 hours DIARRHEAL: 24 hours	• Cook food according to the proper time and temperature guidelines • Make sure to use proper cooling techniques
Clostridium perfringens	• Soil • Animal and human waste • Dust • Insects	• Stews • Casseroles • Meat pies	• Diarrhea • Abdominal pain	8 to 22 hours	24 hours	• Pay close attention to time and temperature guidelines • Carefully wash raw produce • Ensure food preparation equipment is properly cleaned
Shiga toxin-producing E. coli	• Infected workers • Contaminated water	• Undercooked beef • Unpasteurized milk or juice • Contaminated produce	• Severe abdominal cramping • Sudden onset of watery diarrhea, frequently bloody • Sometimes vomiting and a low-grade fever	3 to 5 days	1 to 3 days	• Cook meats and poultry thoroughly • Do not consume raw milk or unpasteurized dairy products • Wash hands after using the bathroom and before preparing or eating food

Bacteria	Sources	Common Food Vehicles	Symptoms	Onset Time	Duration	Specific Controls
Salmonella spp.	• Soil • Sewage • Insects • Feces	• Raw meats • Poultry • Eggs • Milk and dairy products • Fish	• Nausea • Vomiting • Abdominal cramping • Fever	6 to 48 hours	1 to 2 days	• Cook food according to time and temperature guidelines • Prevent cross-contamination of raw meat and ready-to-eat food • Use proper hand-washing techniques
Shigella spp.	• Infected workers • Contaminated water	• Deli salads • Raw vegetables • Milk and dairy products • Poultry	• Vomiting • Abdominal pain • Diarrhea • Blood in stool	12 to 48 hours	5 to 7 days	• Use proper hand-washing techniques • Carefully wash raw produce
Campylobacter jejuni	• Healthy cattle, chickens and birds • Flies • Nonchlorinated water	• Chicken • Raw milk	• Diarrhea • Abdominal pain • Fever • Headache	2 to 5 days	7 to 10 days	• Cook food according to the proper time and temperature guidelines • Prevent cross-contamination of raw meat and ready-to-eat food
Listeria monocytogenes	• Soil • Domestic and feral animals • Raw milk	• Deli meats, hot dogs • Soft cheese • Raw vegetables • Raw meats, poultry and fish	• Flu-like symptoms • Persistent fever	Several days to 3 weeks	Varies	• Cook food according to the proper time and temperature guidelines • Throw away products that have passed their use-by date
Vibrio parahaemolyticus	• Raw and under-cooked fish and shellfish	• Oysters • Clams • Crab • Shrimp • Tuna • Sardines	• Diarrhea • Abdominal cramps • Nausea • Vomiting • Headache • Fever, and chills	4 to 96 hours	3 days	• Purchase seafood from reputable suppliers • Cook seafood according to the proper time and temperature guidelines

VIRUSES

LESSON 3

After completing this lesson, you should be able to:

- Define virus
- Describe means of virus transfer and reproduction
- Identify types of viruses

>TERMS TO KNOW

HEPATITIS A VIRUS A virus primarily found in the feces of infected persons. It is spread from infected food workers to ready-to-eat food including deli meats. It can also be spread to produce and salads and can be found in raw shellfish.

HIV VIRUS A retrovirus spread through blood and bodily fluids. The CDC has found no evidence that the HIV virus can be transmitted through food.

NOROVIRUS Also called viral gastroenteritis, is found in the feces of infected persons. Can also be found in contaminated water.

VIRUSES Microscopic pathogens (smaller than bacteria) that multiply in the living cells of their host.

What Are Viruses?

Viruses are submicroscopic parasites that are smaller than bacteria. Some viruses cause foodborne illnesses. They are usually brought into premises by food handlers who are carriers or on raw food, such as shellfish, which has been in contact with sewage-polluted water. More often, however, viruses hitchhike on another object, known as a vehicle. The main vehicles are hands; hand-contact surfaces, such as refrigerator handles or taps; clothing and equipment; and food-contact surfaces, such as cutting boards. The path along which contaminants are transferred from their sources to food is known as the route of contamination. Most viruses can survive freezing, and they can contaminate both food and water supplies. Viruses are classified as infections.

Virus

Viral Reproduction

Viruses are made up of genetic material in the form of DNA or RNA that is sealed in a protein coat. Viruses do not reproduce on their own, so in order to do so, they must infect a host cell. This factor makes a virus similar to a parasite, which will be discussed later in this chapter. Viruses can enter human cells through a watery pathway between proteins known as a protein channel. Once inside the cell, the virus infects the cell with its genetic information. This information causes the infected cell to start reproducing by making copies of itself. Although all of these cell copies are being created, they are still trapped inside of the original infected cell. They are finally released when, as a defense mechanism, the original infected cell bursts. These cells go on to rapidly infect other previously healthy cells.

Viral Infections

There are two major foodborne illnesses caused by viruses. One is hepatitis A, and the other is norovirus.

Hepatitis A

Hepatitis A is primarily found in the feces of infected persons. It is spread from infected food workers to ready-to-eat food including deli meats. It can also be spread to produce and salads and can be found in raw shellfish.

The most common symptoms associated with hepatitis A are nausea, vomiting, diarrhea, mild fever, headache and general fatigue. After a few days, the infected

Viruses can survive freezing and they can contaminate both food and water supplies.

person may develop jaundice. Symptoms of hepatitis A may be mistaken for flu. Some sufferers, especially children, may exhibit no symptoms at all. Symptoms typically appear two to six weeks after start of infection. They may return over the following 36 months. A vaccine is available that will prevent infection from hepatitis A for at least ten years. To prevent the spread of hepatitis A use proper hand-washing techniques. Also, prohibit infected employees from working on the premises until they have been cleared by a doctor.

Norovirus

Norovirus, also termed viral gastroenteritis, is found in the feces of infected persons. It can also be found in contaminated water. The primary route of contamination is infected food workers who handle food without properly washing their hands after using the restroom. The most common symptoms associated with norovirus are nausea, vomiting, abdominal cramps and diarrhea. The onset time for the infection in humans is usually between 24 and 48 hours. The symptoms can last from 24 to 60 hours. Recovery is usually complete. Prevention methods for norovirus are the same as those for hepatitis A. Make sure workers properly wash their hands, and prohibit infected employees from working on the premises until they have been cleared by a doctor.

Symptoms of Foodborne Illnesses Caused by Viruses

Hepatitis A	Norovirus
Nausea	Nausea
Vomiting	Vomiting
Diarrhea	Abdominal cramps
Mild fever	Diarrhea
Headache	
General fatigue	

HIV Information

It is important to note that the federal Centers for Disease Control (CDC) has a published list of infectious and communicable diseases that can be transmitted through food. The list includes diseases caused by both bacteria and viruses. The CDC has found no evidence that the **HIV virus** can be transmitted through food. Because of this, an employee who has tested HIV positive is not a concern unless that person also suffers from a secondary illness as a result of a foodborne infection or intoxication.

PARASITES AND FUNGI

LESSON 4

After completing this lesson, you should be able to:

- Define parasite
- Describe ways of parasite transmission
- Differentiate between two types of fungi
- Describe hazards and growth conditions of mold
- Identify characteristics of yeasts

>TERMS TO KNOW

AFLATOXINS Toxins produced by mold that can cause serious illness and cannot be killed by cooking.

FUNGI Biological contaminants that can be found naturally in air, plants, soil and water. Fungi can be small, single-celled organisms or larger multi-cellular organisms.

MOLD Microscopic chlorophyll-free fungi that produce threadlike filaments; they can be black, white or of various colors.

PARASITE An organism that lives and feeds in or on another living creature, known as a host, in a way that benefits the parasite and disadvantages the host. In some cases, the host eventually dies.

YEASTS Single-celled microscopic fungi that reproduce by budding and grow rapidly on certain foodstuffs, especially those containing sugar.

Parasites

Parasites are plants or animals that live on or in another plant or animal—the host—to survive. Parasites range in size from tiny, single-celled organisms to worms visible to the naked eye. More and more frequently, they are being identified as causes of foodborne illness in the United States.

Parasites live and reproduce within the tissues and organs of infected human and animal hosts and are often excreted in feces. Parasites may be transmitted through consumption of contaminated food and water.

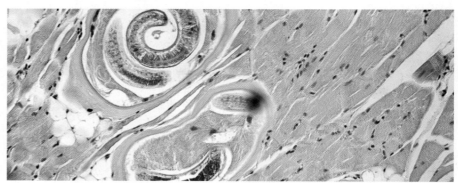

Parasite

Fungi

Fungi, including molds and yeasts, are biological contaminants that can be found naturally in air, plants, soil and water. Some fungi are useful to humans. For example, we use *Saccharomyces cerevisiae* or baker's yeast, to make bread rise and to brew beer. Fungi also aid in the decomposition of dead plants and animals. With respect to food safety, the two types of fungi to be most concerned about are molds and yeasts.

Molds

Molds are threadlike organisms that produce spores. These molds are multi-cellular and can sometimes be seen with the naked eye. It's the spores that give molds their visible color.

Some molds can be dangerous, as they can cause allergic reactions and respiratory problems. Mold also spoils food. Be aware that freezing can prevent or reduce the growth of mold, but it does not destroy it. In some cases, mold can also survive cooking. For example, when you are baking bread, it is likely that some spores will survive in the flour.

Mold grows well under most any conditions, but it grows especially well in food with high acidity and low water activity. It can grow at refrigerated temperatures below

41°F. Mold also grows in bread if the bread is stored under moist conditions after cooling, or if it sits for too long. You'll be able to distinguish mold growth on the surface of food items by its fuzzy appearance. Mold growth on old, spoiled bread will appear green or black in color.

Mold is mostly a spoilage organism, but some molds produce mycotoxins, such as **aflatoxins**, that can cause serious illness. Aflatoxins resulting from mold growth can develop in peanuts or other crops that are stored in a moist environment. Moldy peanuts or any other crops should be disposed of and not eaten. Serious illness can result. Aflatoxins cannot be killed by cooking.

Throw out all moldy food found on your premises unless the mold is a natural part of the food item, as is the case with certain types of cheeses. The Food and Drug Administration (FDA) recommends that you cut away moldy areas in other types of hard cheeses one inch from where they occur.

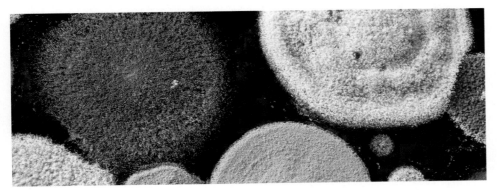

Mold

Yeasts

Yeasts can spoil food at a very quick pace. However, yeasts require oxygen to grow, and when there's no more oxygen, yeasts stop growing. As yeasts grow in food, carbon dioxide and alcohol are produced. Therefore, yeast spoilage can result in a distinct taste or smell of alcohol. Yeasts can appear pink in color, may be slimy and may bubble. Yeasts are sensitive to normal cooking temperatures and are destroyed in this manner.

Yeasts are used in the manufacture of foods such as bread, beer and vinegar; however, they do cause spoilage of foods such as jam, fruit juice, honey, meats and wines. If any of these foods appear to be spoiled by yeast, they should be discarded. Yeasts are very rarely implicated in causing illness, but they will often lead to customer complaints.

Yeast

Types of Parasites

Bacteria	Sources	Common Food Vehicles	Symptoms	Onset Time	Duration	Specific Controls
Anisakis simplex	• Raw or under-cooked seafood	• Sushi • Squid	• Tingling sensation in throat • Cough up nematode • Abdominal pain	1 hour to 2 weeks	Varies	• Cook seafood properly • Freeze food
Giardia lamblia / Giardia duodenalis	• Unwashed hands • Contaminated water	• Raw vegetables	• Nausea • Abdominal cramping • Vomiting	3 to 25 days	1 to 2 weeks	• Use properly treated water • Follow proper hand-washing techniques
Cryptosporidium parvum	• Contaminated soil, food, water or surfaces	• Raw fruits and vegetables	• Watery diarrhea • Coughing • Low-grade fever	2 to 10 days	2 to 4 days	• Use properly treated water • Follow proper hand-washing techniques • Thoroughly wash all vegetables and fruits in clean water
Cyclospora cayetanensis	• Contaminated water • Various types of produce	• Raw fruits and vegetables	• Watery diarrhea • Constipation • Nausea • Abdominal cramping • Fever	1 week	Several days to 1 month	• Use properly treated water • Follow proper hand-washing techniques • Thoroughly wash all vegetables and fruits in clean water
Trichinella spiralis	• Raw or under-cooked food	• Wild game • Pork	• Nausea • Vomiting • Diarrhea • Fever • Abdominal pain • Headache • Eye swelling • Aching joints • Muscle weakness	1 to 2 days	2 to 8 weeks	• Cook all pork and wild game properly

OTHER BIOLOGICAL CONTAMINANTS

LESSON 5

After completing this lesson, you should be able to:

- Identify poisonous plants
- Identify sources of fish poisoning
- Describe the source and symptoms of shellfish poisoning

> TERMS TO KNOW

SCOMBRID A member of a family of fish, including tuna, mahi mahi and mackerel, that can produce histamine toxins when temperature abused.

Poisonous Plants

Plants responsible for causing acute food poisoning include deadly nightshade, death cap mushrooms, daffodil bulbs and rhubarb leaves. Red kidney beans or fava beans occasionally result in food poisoning when consumed raw or undercooked. Common symptoms of illness caused by toxic plants include nausea, vomiting and abdominal pain, likely with an onset time of one to six hours.

Ciguatera Toxin

The gonads, liver and intestines of some fish are highly toxic and may result in food poisoning. In the south Florida, Bahamian, and Caribbean regions, snapper, grouper and mackerel are examples of fish that produce toxins. Symptoms of food poisoning by natural fish toxins include tingling of the fingers, disturbance of vision, paralysis, nausea, vomiting and diarrhea. To prevent illnesses from occurring purchase fish only from reputable suppliers. When the fish is delivered, check that its temperature is 41°F (5°C) or lower. Also, never accept deliveries that show signs of refreezing such as ice crystals or freezer burn.

Scombrotoxic Fish Poisoning

Scombrotoxic fish poisoning is caused by toxins that accumulate in the body of **scombrid** fish during storage, such as tuna, mackerel, sardines, pilchards, herring, anchovies and salmon. It is likely to occur when temperatures are above 39.2°F (4°C). Symptoms include headache, nausea, vomiting, abdominal pain, a rash on the face and neck, a burning or peppery sensation in the mouth, sweating and diarrhea. These symptoms may last up to eight hours. Again, purchase fish from only reputable suppliers. Also check for signs of time/temperature abuse.

Shellfish Toxins

Shellfish food poisoning may result from the consumption of mussels and other bivalves that have fed on poisonous algae or plankton. These toxins have no odor and no taste and cannot be destroyed by freezing or cooking. Symptoms caused by shellfish poisoning include headache, a floating feeling, dizziness, lack of coordination and tingling of the mouth, arms or legs. The best way to protect yourself from allowing shellfish food poisoning to make its way onto your premises is to purchase shellfish only from reputable suppliers. Seafood certification tags, which list where and when the shellfish were harvested, are required on all shellfish and must be kept on file for at least 90 days.

Other Biological Contaminants*

Contaminant	Common Food Vehicles	Specific Controls
Scombrotoxin	Primarily associated with tuna fish, mahi-mahi, blue fish, anchovies, bonito, mackerel; also found in cheese	Check temperatures at receiving; store at proper cold holding temperatures; Buyer specifications: obtain verification from supplier that product has not been temperature abused prior to arrival in facility.
Ciguatoxin	Reef fin fish from extreme SE US, Hawaii, and tropical areas; barracuda, jacks, king mackerel, large groupers and snappers	Purchase fish from approved sources. Fish should not be harvested from an area that is subject to an adverse advisory.
Tetrodoxin	Puffer fish (Fugu; Blowfish)	Do not consume these fish.
Mycotoxins *Aflatoxin*	Corn and corn products, peanuts and peanut products, cottonseed, milk and tree nuts such as Brazil nuts, pecans, pistachio nuts and walnuts. Other grains and nuts are susceptible but less prone to contamination.	Check condition at receiving; do not use moldy or decomposed food.
Patulin	Apple juice products	Buyer specification: obtain verification from supplier or avoid the use of rotten apples in juice manufacturing.
Toxic mushroom species	Numerous varieties of wild mushrooms	Do not eat unknown varieties or mushrooms from unapproved source.
Shellfish toxins *Paralytic shellfish poisoning (PSP)*	Molluscan shellfish from NE and NW coastal regions; mackerel, viscera of lobsters and Dungeness, tanner and red rock crabs	Ensure molluscan shellfish are from an approved source and properly tagged and labeled.
Diarrhetic shellfish poisoning (DSP)	Molluscan shellfish in Japan, western Europe, Chile, NZ, eastern Canada	
Neurotoxin shellfish poisoning (NSP)	Molluscan shellfish from Gulf of Mexico	
Amnesic shellfish poisoning (ASP)	Molluscan shellfish from NE and NW coasts of NA; viscera of Dungeness, tanner, red rock crabs and anchovies.	

*Although the 2009 FDA Food Code catagorizes toxins as chemical contaminants, in food safety management, they are traditionally catagorized as biological hazards, so that is how we have decided to present them here.

ASSESSMENT

1. What are the three main types of contamination?
   a. Physical, biological, mineral
   b. Biological, chemical, mineral
   c. Physical, biological, chemical
   d. Botanical, chemical, biological

2. When may food be contaminated?
   a. Before delivery
   b. During preparation by employees with poor hygiene
   c. After it is served to customers
   d. All of the above

3. Which of the following two statements are accurate about bacteria?
   a. The majority of bacteria are harmless
   b. Food spoilage bacteria cause serious illness
   c. Food poisoning bacteria are the most significant biological contaminant
   d. There are a large number of bacteria that spoil food

4. Which of these is true about spores produced by bacteria?
   a. Spores are activated by the thawing process
   b. Spores need food to survive
   c. Spores need water to survive
   d. Spores are very difficult to kill

5. The acronym FAT TOM helps us to remember the variables that affect bacterial growth. What do the two "T"s stand for?
   a. Total
   b. Time
   c. Thermometer
   d. Transmit
   e. Temperature

6. At what temperature do food poisoning bacteria most rapidly multiply (the Temperature Danger Zone)?
   a. Below 41°F
   b. Between 41°F and 68°F
   c. Between 41°F and 135°F
   d. Above 135°F

7. Which bacteria cause a highly deadly foodborne illness?
   a. *Staphylococcus aureus*
   b. *Clostridium botulinum*
   c. Shiga toxin-producing *E. coli*
   d. *Salmonella* spp.

8. Which of these is a main vehicle for virus transfer?
   a. Clothing and equipment
   b. Hand-contact surfaces
   c. Food-contact surfaces
   d. Flies
   e. a, b and c

9. Which of these is NOT true with respect to mold?
   a. Mold can grow at refrigerated temperatures below 41°F
   b. Molds can cause allergic reactions and respiratory problems
   c. Molds can spoil foods
   d. Cooking always destroys molds

10. Which of these two statements are true concerning shellfish food poisoning?
    a. The toxins have no odor and no taste
    b. Illness sometimes results in death
    c. Toxins accumulate in the shellfish during storage
    d. The toxins can be destroyed by freezing or cooking

OTHER SOURCES
OF CONTAMINATION

CHEMICAL CONTAMINATION

LESSON 1

After completing this lesson, you should be able to:

- Identify the most common chemical contaminants
- Understand the importance of proper food preparation equipment
- Prevent chemical contamination

>TERMS TO KNOW

CHEMICAL CONTAMINATION
The contamination of food by chemical substances such as pesticides and cleaning solutions.

What Is Chemical Contamination?

Chemical contamination is the presence of unwanted chemical components in food or the food environment. Chemical contamination may be as straightforward as pesticide poisoning from improperly cleaned produce, or as subtle as toxic metal poisoning from improper food preparation or storage equipment.

Chemicals

Chemicals such as pesticides, cleaning agents and refrigerants are commonly used in food preparation areas and are a common source of chemical contamination.

First and foremost, when using any chemical follow all EPA-registered label use instructions. This includes strictly following any warnings as well as proper chemical mixing. Never use a stronger chemical than is needed to achieve the desired result. Food should also be stored, covered or otherwise protected when using chemicals. Because of the difficulty of protecting food during operating hours, a good rule of thumb is to try to use toxic chemicals, such as pesticides, after operating hours when all exposed food is properly stored. In addition, consider using a licensed professional to apply chemicals such as pesticides.

To prevent chemical contamination through improper storage, use the following guidelines. First, chemicals should have a separate storage area—never store them with food, utensils or other food preparation equipment. Second, store chemicals in their original EPA-registered label use containers along with instructions for use. If you must transfer chemicals to another container, ensure that container is properly labeled and stored. Finally, any containers or utensils used with chemicals should never be used for food storage or preparation.

Food Storage

Toxic metal poisoning can occur when acidic foods such as tomatoes are stored or cooked in aluminum or copper containers. The acid in these foods leaches the toxic metals from the container into the food.

Examples of other foods with high acid levels that can create this problem are grapefruit, mayonnaise, vinegar, oranges and pineapples. Toxic metal poisoning causes both unpleasant tasting food and illness. To prevent this type of poisoning, never store food in galvanized metal. Take steps to ensure that all storage and preparation equipment, including utensils, are food grade.

Equipment

Toxic metal poisoning can also occur through improper installation or maintenance of carbonated-beverage dispenser systems.

This type of toxic metal poisoning occurs if carbonated water flows back through copper supply lines. As occurs with acidic foods, the carbonated water will leach copper from the lines and contaminate the beverage. To prevent carbonated beverages from becoming contaminated, you should ensure that all beverage systems are professionally installed and maintained and fitted with a backflow prevention device.

Avoid storing or cooking acidic foods in copper containers.

Other Sources of Chemical Contamination

Suppliers

To prevent chemical contamination that may come through your food supply, ensure that your suppliers are following all appropriate regulations for raising, slaughtering and harvesting their product. This may include random quality checks and appropriate penalties.

Lubricants

Ensure that any lubricants or oils used with kitchen equipment are food grade.

Sanitization

Ensure that your staff are properly trained in cleaning and sanitizing practices. This includes properly washing all appropriate foods as well as thoroughly washing their own hands after using any chemicals.

Make sure your staff are properly trained in washing foods.

PHYSICAL CONTAMINATION

LESSON 2

After completing this lesson, you should be able to:

- Identify common sources of physical contamination
- Show how physical contamination often occurs
- Discuss how to prevent physical contamination

>TERMS TO KNOW

PHYSICAL CONTAMINATION
Occurs when any foreign object becomes mixed with food and presents a hazard or nuisance to those consuming it.

What Is Physical Contamination?

Physical contamination occurs when any foreign object becomes mixed with food and presents a hazard or nuisance to those consuming it.

Foreign objects can range from a piece of hair, to naturally occurring food objects such as bones or stalks. Examples of common foreign objects that present physical-contamination risk are:

- Dirt
- Hair, skin, scabs or fingernails
- Pencils, pens, ink, etc.
- Jewelry
- Glass, metal fragments, wood and paint chips
- Paper fragments
- Plastic and other food packaging items such as twist ties and staples
- Dead insects, rodents or rodent droppings

Physical contamination can occur at any stage in the food creation, delivery or preparation process.

Prevent Contamination

Delivery

As a first step to preventing physical contamination, carefully inspect all food deliveries to ensure that no physical contamination has occurred prior to delivery. Warning signs that food has been compromised can include damaged packaging and food that is not properly protected or presented. In addition, carefully screen all suppliers and ensure their food production facilities meet appropriate standards. Don't hesitate to replace a supplier if they repeatedly deliver food in poor condition.

Storage

Once food has been delivered, make sure that it is properly covered and stored at your location. Regularly inspect food storage areas for cleanliness and signs of pest problems. Also make sure storage areas are well maintained. For example, if a light-bulb is broken in a storage area, ensure that the glass is immediately cleaned up and all food in the area is inspected to verify that it hasn't been contaminated with glass.

Preparation

Food is probably most vulnerable when it is being prepared for consumption. First, employees are removing food from storage and placing it on food preparation workstations. Second, employees are touching the food with their hands and exposing food to any contaminants they may have on their person such as jewelry or hair. And finally, food preparation equipment and utensils can themselves present a source of contamination if not adequately cleaned, sanitized, maintained and replaced when damaged.

To minimize the likelihood that food will be contaminated during preparation, ensure that good food safety policies are in place and that employees are well trained in these policies. Examples include:

- Cleaning and inspecting workstations prior to beginning food preparation
- Thorough hand washing prior to handling food
- Ensuring that all employees wear hair nets and, if applicable, beard nets
- Not allowing jewelry in the food preparation area
- Not allowing loose items in shirt pockets such as pens, pencils, paper or coins

Maintenance

Good maintenance practices can also minimize the possibility of accidental physical contamination. Regularly inspect utensils and other food preparation equipment. Replace those that are failing or might otherwise present a contamination risk. In addition, ensure a prompt and thorough cleanup after any mechanical repairs. This reduces the risk that any loose parts or waste materials such as screws or drill shavings might pose a contamination threat.

Customers

If you are using food displays, customers themselves can pose a physical-contamination risk. To minimize this risk, ensure that display cabinets are kept clean and provide a good barrier between the food and the customer. Also, inspect all food from the display case prior to delivering it to the customer. Finally, do not allow customers or any other unauthorized persons to enter any food preparation area, including display cases.

INTENTIONAL CONTAMINATION

LESSON 3

After completing this lesson, you should be able to:

- Discuss the importance of a good food safety program
- Identify common points of vulnerability
- Address common points of vulnerability

It is an unfortunate reality that in today's world, food managers must also prevent the purposeful contamination of food. Intentional contamination of food can be a goal not just of activist groups and terrorist organizations, but also of employees, former employees and competitors.

A Good Food Safety Program

As with other forms of food contamination, intentional contamination can occur at any stage in the food creation, delivery or preparation process. To prevent intentional contamination of food, food safety must be taken as seriously as safety of funds, people and equipment. Adequate food safety requires that each food establishment have its own comprehensive approach to food safety. While it is challenging to set up policies, procedures and training for a comprehensive food safety program, remember that a single instance of intentional contamination can be catastrophic for your business.

A good food safety program can make it difficult for any intentional contamination to take place. To be comprehensive, a good food safety program must take into account the most common areas of food vulnerability. This includes the people element, as well as the building element. In addition, when considering the building element, both interior and exterior vulnerabilities are important.

A good food safety program must take into account the most common areas of food vulnerability.

People

Employees

Let's consider the human aspect of intentional contamination. First, never allow anyone but on-duty employees in food preparation and storage areas. Limit on-duty employees to "essential items only" while at work. Consider a two-employee rule for food preparation areas. Carefully monitor food preparation areas. In addition, carefully screen employees prior to their hire. While it can be a tedious part of the employment process, identity verification and reference checks can help you keep a safe food and work environment.

Suppliers

It is just as important to carefully screen and approve suppliers as it is employees. Ensure that all approved suppliers are clearly identified. Inspect and document all deliveries, and never allow any deliveries from any nonauthorized vendors. Ask suppliers about their food safety programs and whether tamper-evident packaging is available.

Building

Interior

On the inside of your establishment, control the entrances and exits for employee-only areas such as food preparation areas, food display areas and kitchens. Use good lighting and cameras to eliminate hiding areas in all parts of the building. Be sure that if you have self-service areas such as salad bars, these areas are carefully monitored.

Exterior

Ensure that the outside of your establishment also meets good safety standards, including a well-lit exterior, locked back doors and a ventilation system that can be accessed only by authorized personnel. Ensure that no unauthorized individual, even employees, can access the building after normal business hours.

Policies and Training

As you can see, a good food safety program addresses vulnerabilities from people, as well as basic facilities safety. However, the best food safety program is useless if your employees are unaware of it or don't follow it. It is vital that all employees be thoroughly trained in your food safety practices and be required to follow them. In addition, employees should be trained and encouraged to report any suspicious activity that may be a sign of intentional food tampering.

FOOD ALLERGENS AS CONTAMINANTS

LESSON 4

After completing this lesson, you should be able to:

- Explain how food allergies affect the body
- Identify the most common causes of food allergies
- List good practices that can help prevent allergic reactions

The Food Allergen Labeling and Consumer Protection Act (FALCPA) took effect in 2006. This law requires all food containing a major food allergen, or a food ingredient that contains protein derived from a food allergen, to be clearly labeled.

Nearly seven million Americans suffer from food allergies. Food allergies affect 2% of adults and 5% of infants and children in the United States. Food allergies can be deadly, so proper labeling is an important element of food safety.

Food Allergy Symptoms

A **food allergy** is the body's immune system responding to a food that it mistakenly believes is harmful. Depending on the person, the immune response could be immediate and severe or delayed for a period of time with milder symptoms. Symptoms of food allergies include a tingling sensation in the mouth or throat, itching in and around the mouth, face and/or scalp, swelling, including swelling of the tongue, throat, face, eyes, hands and feet, difficulty breathing, including wheezing or shortness of breath, rash or hives, nausea and/or vomiting, abdominal cramps and diarrhea and loss of consciousness.

Food Allergy Symptoms
Tingling sensation in the mouth or throat
Itching in and around the mouth, face and/or scalp
Swelling
Difficulty breathing
Rash or hives
Nausea and/or vomiting
Abdominal cramps and diarrhea
Loss of consciousness

A serious allergic reaction that is rapid in onset and life-threatening is called **anaphylactic reaction**. Food allergies are believed to be the leading cause of anaphylaxis outside of a hospital setting. Because food allergies can cause death, it is important to be aware of and train your staff to recognize an allergic reaction so that proper medical treatment can be administered.

Food allergies cannot be cured. The only way to prevent an allergic reaction is to avoid the food that triggers it.

ANAPHYLACTIC REACTION A severe allergic reaction affecting the whole body, often within minutes of eating the food, which may result in death.

FOOD ALLERGY An identifiable immunological response to food or food additives, which may involve the respiratory system, the gastrointestinal tract, the skin or the central nervous system. The most common food allergies are caused by nuts (especially peanuts), eggs, wheat, shellfish, milk and soy.

Common Food Allergens

Eight major foods account for 90% of all food allergies in the United States: milk, eggs, fish, shellfish, tree nuts, peanuts, wheat and soy. These major allergens must be properly labeled under the 2006 Food Allergen Labeling and Consumer Protection Act.

The 2006 Food Allergen Labeling and Consumer Protection Act applies only to packaged, FDA regulated foods. The labeling requirement does not cover food service and retail establishments. Food placed in a wrapper or container in response to a customer order (such as a box for a sandwich) does not have to be labeled.

In addition to naturally occurring food allergens, food additives can also bring about life-threatening allergic reactions. Ensure that food additives are used in strict compliance with the instructions provided by the manufacturer and that they are properly labeled. MSG, or monosodium glutamate, is a food additive that can cause allergic reactions in some people. Symptoms tend to occur within one hour of eating three grams or more on an empty stomach. Severe, poorly controlled asthma may predispose someone to a reaction to MSG.

Major Allergens

Milk   Eggs   Fish and shellfish   Tree nuts and peanuts   Wheat   Soy/soybeans

Preventing Allergic Reactions

As with many other elements of food safety, implementing appropriate policies and training can go a long way to preventing allergic reactions. First, train your staff to know which menu items contain major food allergens and to recognize the symptoms of a reaction. Ensure that at least one person on each shift is thoroughly familiar with all the food ingredients you use. Train your cooking staff to be able to prepare menu items that are allergen-free if a customer requests it. Impress upon your staff that people suffering from food allergies must totally avoid the food in question, as even tiny amounts can cause a severe reaction. Use good food preparation practices, and avoid cross-contamination with an allergenic food.

Other good practices for preventing allergic reactions include:

- Carefully reading all labels
- Labeling allergenic items on the menu
- Using a separate prep area for items containing allergenic ingredients
- Carefully cleaning tables and utensils that have come into contact with an allergen
- Serving sauces on the side
- Avoiding product substitutions for menu items
- Being alert if a customer indicates they have an allergy, having an emergency procedure in place to handle allergic reactions and calling for treatment when necessary

ASSESSMENT

1. All of the following are common causes of chemical contamination except for:

   a. Improperly labeled cleaning supplies
   b. Improperly cleaned produce
   c. Improper food storage
   d. Improperly canned food
   e. Improper use of pesticides

2. What type of container should NOT be used to store acidic foods?

   a. Metallic
   b. Plastic
   c. Wood
   d. Glass

3. In which two ways can a food handler chemically contaminate food?

   a. Not washing floors properly
   b. Not washing hands properly
   c. Not washing food properly
   d. Not washing walls properly

4. What two warning signs should you look for to identify contamination at delivery?

   a. Itemized list of the food being delivered
   b. Food stored at a proper temperature
   c. Food not properly protected
   d. Damaged packaging

5. Each of the following factors makes food preparation potentially hazardous except:

   a. Moving food from storage to workstations
   b. Cleaning the food preparation area too often
   c. Employees touching the food
   d. Using improperly cleaned and sanitized equipment and utensils

6. What is the main reason for having a good food safety program?

   a. To market yourself to potential customers
   b. To prevent contamination of food
   c. To attract top quality suppliers
   d. To maintain reliable employees

7. What areas of your building should you maintain as part of your food safety program?

   a. Entrances and exits
   b. The interior
   c. The exterior
   d. All of the above

8. What areas of vulnerability does a good food safety program address?

   a. People and buildings
   b. Weather
   c. FDA Food Code
   d. Hours of operation

9. What is a food allergy?

   a. A reaction to a food that a person does not like
   b. A curable condition that is easily treated over the counter
   c. The body's immune response to a food it mistakes as harmful
   d. An avoidance of a certain color of food

10. All of the following are common allergens except:

    a. Milk
    b. Eggs
    c. Rice
    d. Peanuts

HANDLING
FOOD SAFELY

THE FOOD HANDLER

LESSON 1

After completing this lesson, you should be able to:

- Explain personal-hygiene control measures
- Describe how these measures can prevent food safety hazards

People, especially food handlers, are potentially the most important hazard in a food premises. They are sources of physical, chemical and biological hazards and can cause cross-contamination as a result of poor hygiene practices.

Setting Personal-Hygiene Standards

Good personal-hygiene habits will reduce or eliminate the principal contamination hazards associated with people.

Your basic personal-hygiene program should consist of the following directives:

- Employees must practice proper hand-washing techniques and employ proper glove use, when applicable.
- Employees must maintain a high level of personal cleanliness.
- Employees must wear proper work attire.

Preventing Contamination

Let's look at ways that physical contamination can be controlled or eliminated. The principal hazards associated with the human body are bacterial. Bacteria that can be reduced by proper hand washing include *E. coli* and *Shigella* **spp**. Another big concern is contamination from the bacteria *Staphylococcus aureus*. The bacteria are often present in boils, in skin infections and cuts, on hands, in the nose, mouth and ears and on hair.

When it comes to boils, skin infections and cuts, make sure to:

- Cover all wounds with a waterproof dressing, preferably blue. (The dressings should be blue so that they can easily be seen if they fall into food.)
- Exclude staff members from handling time/temperature controlled for safety (potentially hazardous) food if they have boils, skin lesions or infections.

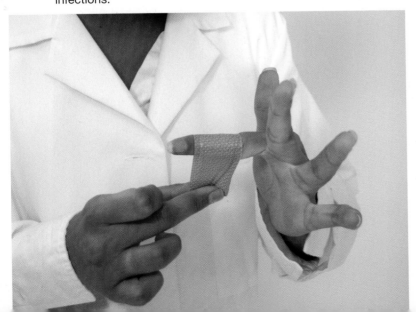

E. COLI A bacteria found in the intestines of mammals. It can be found in ground beef and also contaminated produce.

STAPHYLOCOCCUS AUREUS A bacteria commonly found on the skin, nose and hands of one out of two people. It is transferred easily from humans to food when people carrying the bacteria handle the food without washing their hands. This bacteria also produces toxins that multiply rapidly in room-temperature food.

***SHIGELLA* SPP.** A bacteria found in the feces of people with shigellosis. It can be found in ready-to-eat foods such as greens, milk products and vegetables and also in contaminated water. The most common method of transmission is cross-contamination. Flies can also be carriers of this type of bacteria.

Hands

Because hands come into direct contact with food during preparation, they often act as a vehicle, transferring bacteria from other items or from other parts of the body directly onto the food itself.

When it comes to hands, employees must:

- Keep hands clean at all times.
- Keep nails short and clean.
- Not use false nails or nail polish.

Nose, Mouth and Ears

When it comes to the nose, mouth and ears, employees must adhere to the following rules:

- Do not cough or sneeze over food or equipment.
- Do not bite nails, blow noses or scratch ears or skin.
- Do not lick fingers, blow onto food or equipment or spit.
- Do not snack or chew gum.
- Do not use a finger to taste the food. Employees should always use a clean spoon to taste food.

Hair

When it comes to hair, insist that employees:

- Wear hairnets or protective hats, or even hairnets under protective hats when possible.
- Do not touch or comb their hair while at work.
- Do not scratch their heads.

EMPLOYEE HEALTH

LESSON 2

After completing this lesson, you should be able to:

- Set a personal-hygiene policy
- Describe how and when you must restrict employees from work and when employees can return to work

Personal-Hygiene Policy

It's important that you set a personal-hygiene policy that is clear and easy to follow. The policy MUST outline the actions that you take, as well as the actions your employees must take when they are sick. It's important that sick employees notify you before they come to work or notify you immediately if they become ill while at work.

It's also vital that you remain approachable so that employees feel comfortable informing you when they are ill.

Restricting Employees from Work

Food handlers are particularly hazardous when they are ill. You must ensure that everyone on your staff knows they need to report any illnesses to you promptly. As a manager, there will be times when you must restrict employees from working with food, working in food contact areas and sometimes even from working within the establishment. For example, restrict an employee from working with food or in food contact areas if he has a sore throat with fever. If you serve a **high-risk population** (such as the elderly, the very young, people who are immunocompromised, pregnant women and allergen-sensitive people), you must exclude the employee from the establishment. Exclude any employee from the establishment who has vomiting, diarrhea or jaundice.

High-risk populations

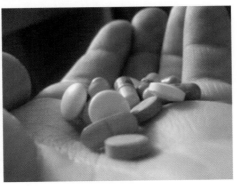

Employees are required to notify management if they are experiencing symptoms that may increase the risk of infecting high-risk clientele.

Reporting Symptoms

Employees must notify you if they are experiencing any of the following symptoms:

- Vomiting
- Diarrhea
- Jaundice
- Sore throat with fever
- A wound or lesion, such as a boil or infected wound, that is open or draining and cannot be protected by an impermeable, proper-fitting cover or bandage

Reporting Diseases

In addition to ensuring that employees notify you when they are sick, you need to understand when an employee illness requires notification to a regulatory agency. It's your responsibility to inform the appropriate regulatory authority when an employee is either jaundiced or diagnosed with any of the following illnesses:

- Norovirus
- Hepatitis A virus
- *Shigella* spp.
- Shiga toxin-producing *E. coli*
- *Salmonella typhi*

An employee must also notify you if he has had a prior illness due to *Salmonella typhi,* within the past three months, without receiving antibiotic therapy.

If an employee has been excluded from work due to an illness or disease that required regulatory notification, you must work with that employee's health care practitioner and/or the regulatory agency to determine when the employee may return to work. No matter the cause, whenever an employee previously restricted from work returns to work, it is critical that good hygiene is observed, particularly hand washing, at all times.

Remember, however, that sometimes food handlers may be healthy carriers. It is therefore essential that you and your staff observe the highest standards of personal hygiene at all times.

PERSONAL HYGIENE OF EMPLOYEES

LESSON 3

After completing this lesson, you should be able to:

- Explain the proper kinds of clothing, aprons and hair restraints
- Describe how smoking can affect the food in your environment

>TERMS TO KNOW

PERSONAL HYGIENE Standards of personal cleanliness habits, including keeping hands, hair, and body clean and wearing clean clothing in the food establishment.

It is imperative that your employees—especially food handlers—maintain a high level of **personal hygiene**, as this is crucial to staying healthy and keeping food safe.

Employees should be in the practice of showering or bathing on a daily basis and should also make a habit of frequent, proper hand washing.

Clothing

Food handlers must always wear clean and washable protective clothing. Protective garments should be appropriate for the work being carried out. Protective garments should also completely cover ordinary clothing.

When dressing, it is important to dress from the top down. First, put on the hairnet or the hat. Then, put on the other pieces of clothing. And finally, put on the shoes. This way, you will avoid contaminating the protective clothing with bacteria from your hair and work shoes.

Remember, the main purpose of protective clothing is to protect food from contamination.

Whenever possible, food handlers should change into their work clothes once they've entered the work premises.

Aprons

Always begin the work day with a fresh, clean apron. Food handlers must remove aprons upon leaving the food preparation areas. There should be a designated area for aprons to be hung or stored, for example, when a food handler has to remove his apron and take out some trash or visit the toilet. If an apron becomes soiled, it should be changed immediately.

Shoes

Suitable footwear should be worn to prevent slipping and to protect the feet. Appropriate footwear include clean close-toed shoes with a low heel and nonskid soles.

Jewelry and Perfume

Other potential sources of contamination by food handlers are jewelry and perfume.

Food handlers should not wear earrings, watches or rings, as they may hide dirt and bacteria. Stones and small pieces of metal present in jewelry may physically contaminate food, as can nail polish or false nails.

Finally, avoid the use of strong-smelling perfume or aftershave as they may taint the food.

Smoking

It is illegal to expose food to the risk of contamination by smoking in food preparation areas or while handling food.

- People touch their lips and can transfer bacteria to food from their mouth.
- Cigarettes contaminated with saliva may be placed on work surfaces.
- Smoking encourages coughing.
- Cigarette butts and ash may land on and contaminate food.

Food handlers should smoke only in designated areas.

In many areas of the U.S., recent legislation has banned smoking from taking place anywhere inside the premises. If this is the case in your area, you should designate an area outside of the premises where smoking is permitted.

Policies

It is always good practice to check with local regulatory agencies regarding the requirements for proper work attire and policies on smoking, eating and drinking.

If such requirements exist, make sure you have written policies to make current and potential employees aware of what is or will be expected of them when reporting to work.

HAND WASHING

LESSON 4

After completing this lesson, you should be able to:

- Demonstrate how staff should wash their hands
- Explain why staff should wash their hands
- Describe when staff should wash their hands

>TERMS TO KNOW

HAND WASHING The process of cleansing the hands with soap and water to thoroughly remove soil and/or microorganisms. Food workers must clean their hands up to their elbows.

Hand washing is one of the most important actions that can be taken to prevent the spread of foodborne illnesses.

Manager Responsibility

As the manager, it is your responsibility to train your employees in the proper hand-washing technique. You must also monitor that they put it into practice.

You can ensure that your staff adheres to these good practices by demonstrating the hand-washing technique, placing posters in the food area to remind employees of the proper technique and, of course, following these guidelines yourself.

Why Hands Need to Be Washed

The main objective of washing hands is to reduce the number of pathogens on hands to a safe level. A salad employee who carries the *Shigella* bacteria and does not wash hands properly can pass the bacteria to customers through the salad.

A single-wash procedure is normally sufficient. However, activities likely to result in a large number of pathogens on hands—such as using the toilet or cleaning up vomit or animal feces—should be followed by a double wash instead.

The double-wash procedure includes an extra, initial stage to brush the fingernails and fingertips under running water, using liquid soap on the nailbrush.

When to Wash Hands

Food handlers must wash their hands before starting work and should also wash them regularly throughout the work day, especially following certain activities. Hands should always be washed after using the toilet, without exception.

Hands must also be washed:
Upon entering a food preparation or serving area
Before handling TCS potentially hazardous food or raw food
After clearing tables or bussing dirty dishes
After touching or taking out the trash
After dealing with an ill customer or co-worker
After putting on or changing a dressing covering things like boils, skin infections and cuts
After cleaning animal feces, handling boxes contaminated with bird droppings or handling a baby's diaper
After handling hazardous chemicals
After touching the hair or face
After eating or smoking
After coughing, sneezing or blowing the nose
After handling money

How to Wash Hands

To reduce bacteria to a safe level through hand washing, it is essential that the correct procedure is followed. Hands should be washed for a total of 20 seconds, including rubbing the hands together for 10 to 15 seconds as in step 3.

1. Wet hands and exposed portions of the arms with clean, running water that is as hot as you can comfortably stand. The temperature should be at least 100°F (38°C).

2. Apply a liquid soap.

3. Rub the hands together vigorously, cleaning all parts of the hands and arms, especially the fingertips and around the nails. Do this for 10 to 15 seconds.

4. Rinse hands and arms.

5. Completely dry hands and arms using a single-use paper towel or an air dryer.

If the hands are likely to be heavily contaminated, for example, after going to the toilet, changing a dressing, or cleaning up feces or vomit, a nailbrush should be used before the normal wash. The nails should be rubbed gently on a clean nailbrush under running water. This is known as the double wash procedure.

Where to Wash Hands

In order to avoid contamination, hand washing must take place in a hand-washing specific basin, such as a dedicated hand-washing sink or an approved automated hand-washing facility. Hand washing should not occur in sinks used for preparing food or in a service sink.

Bare-Hand Contact

Bare-hand contact with ready-to-eat food should be avoided whenever possible. In fact, the current FDA *Food Code* prohibits bare-hand contact from taking place on food service premises. Employees may not have bare-hand contact with ready-to-eat foods in an establishment that serves a highly susceptible population. This does not pertain to ready-to-eat food that is being added as an ingredient to food that will be appropriately cooked. If contact with ready-to-eat food is necessary, have the appropriate utensils available to employees, such as:

- Deli tissue
- Spatulas
- Tongs
- Scoops
- Single-use gloves
- Dispensing equipment

Hand washing should occur in a dedicated hand-washing sink.

In some cases, approval to touch ready-to-eat food with the bare hands may be obtained by contacting the proper regulatory authority.

Special requirements

Prior to allowing employees bare-hand contact, the employees must sign that they have received training in the following areas:

- The risks associated with contacting the specific ready-to-eat food items with their bare hands
- Proper hand-washing methods, including when to wash hands and where to wash hands
- Proper fingernail maintenance, including the proper use of a fingernail brush
- Prohibitions against jewelry, such as rings, where bacteria and germs can evade the effects of hand washing

General good hygiene practices

Employee health policies, such as what to do in cases of illness, when illness will exclude employees from bare-hand contact or the workplace altogether, and when illness might result in restrictions to bare-hand contact or food contact in general.

Additionally, if bare-hand contact is to be allowed, the food service establishment must utilize two or more of the following additional safeguards:

- Double hand washing before contact.
- The use of nail brushes during hand washing.
- The use of hand sanitizer following proper hand washing.
- An incentive program that encourages and/or assists food employees to refrain from working when ill or otherwise contagious. This could include paid time off for documented illness, or reassignment in duties without impacting pay, among others.
- The use of another control measure that has been approved by the regulatory agency with jurisdiction over the food service establishment.

Please note that these safeguards must be utilized in addition to, not instead of, regular and proper hand washing.

The final step required for bare-hand contact of ready-to-eat foods is documentation of corrective actions taken when steps one and/or two are not adhered to. This documentation must include a plan for implementing these corrective actions. It may be included as a part of the food service establishment's HACCP log (described in Chapter 9) or as a stand-alone document, but it must be in written format.

Gloves

In order to maintain a clean and safe environment, employees should wear gloves to inhibit the spread of bacteria.

When used properly, gloves can aid in the service of safe food on your premises by acting as an added layer of protection between the hands and food. Also, when customers see your employees wearing gloves, it tells them that your business is committed to serving safe food and, therefore, committed to protecting their health.

Types of Gloves and Uses

When it comes to purchasing the gloves to be used on your premises, there are several factors to be aware of.

First, gloves used in food service establishments should be single-use gloves only. Do not try to wash and reuse gloves.

Second, buy gloves in a variety of different sizes. If the gloves are too large, your employees will have a hard time keeping them on their hands. If the gloves are too small, it will be easier for them to rip or tear.

Third, do not use latex gloves. It is common for people to be sensitive to latex or to develop allergies due to the extended use of latex gloves. Alternative materials include polyvinyl, nitrile, chloroprene and polyethylene.

Fourth, match the proper type of glove to the appropriate task. Buy looser-fitting, less expensive gloves for times when frequent changing is necessary. Buy more durable, form-fitting gloves that cost a little more for repetitive tasks that require less frequent changing.

How to Use Gloves

Remember, wearing gloves does not replace proper hand washing. In fact, the first step to wearing gloves is to wash and dry your hands using the proper hand-washing technique.

Since gloves essentially act as a second skin, they can be contaminated the same way that hands can. Therefore, gloves should be changed after touching anything that may be a source of contamination.

Gloves should never be worn for more than four hours, as perspiration and bacteria can build up under the gloves. So, even if you have no other reason to, dispose of your gloves, properly wash and dry your hands, and then put on a fresh pair after four hours have passed.

When to change your gloves

When changing tasks

After touching raw meat

Before handling cooked or ready-to-eat food

After covering your mouth when sneezing or coughing

After touching your face or hair

When they become soiled or torn

After four hours

ASSESSMENT

1. What is the most important reason for promoting good personal hygiene?

   a. To increase morale in the workplace

   b. To remove unwanted odors from your establishment

   c. To reduce contamination hazards associated with people

   d. To add another chore for employees

2. What is the most common bacteria associated with people?

   a. *Staphylococcus aureus*

   b. *Cryptosporidium*

   c. *Clostridium botulinum*

   d. *Clostridium perfringens*

3. Why should you use a colored bandage to cover wounds?

   a. It is cheaper for people in the food service industry

   b. It is required in the FDA Food Code

   c. It can be easily seen if it falls into food

   d. It is the only one available on your premises

4. In which two cases should employees inform you of illness?

   a. When they have visited a close family member in the hospital

   b. When they or a household member have diarrhea

   c. When they had chicken pox as a child

   d. When they have eaten a meal that caused food poisoning

5. An employee would NOT have to notify the management of the following symptoms?

   a. An open wound that cannot be protected by a bandage

   b. A sore throat with fever

   c. A migraine headache

   d. Vomiting, diarrhea or jaundice

6. What is the maximum time you should wear a pair of gloves when doing the same task?

   a. 2 hours

   b. 4 hours

   c. 8 hours

   d. 30 minutes

7. When should food handlers wear their aprons?

   a. When taking out the trash

   b. When entering the main part of the restaurant

   c. In the food preparation area

   d. When visiting the restroom

8. What is the most important reason for wearing hair restraints?

   a. They keep hair from falling in the food

   b. They give the customers confidence that food safety standards are being followed

   c. They look fashionable

   d. They are required by the management team of your business

9. Whom should you contact with questions about food safety policy?

   a. Local regulatory agency

   b. FDA

   c. OSHA

   d. Restaurant supply center

10. What is the most effective way to communicate your policies?

    a. Post them on the Internet

    b. Tell your employees daily

    c. Recite them to your customers

    d. Write them down and share them with employees

FROM PURCHASE TO SERVICE

TCS FOODS (PHFs)

LESSON 1

After completing this lesson, you should be able to:

- Identify TCS foods
- Prevent time/temperature abuse
- Prevent cross-contamination between TCS foods and safe foods

What Are TCS Foods?

TCS foods, otherwise known as "time/temperature control for safety foods (potentially hazardous foods)" are food products that—under the right circumstances—support the growth of foodborne illness-causing microorganisms. Specifically, TCS foods (PHFs) are foods that are high in protein, have neutral or slightly acidic pH levels and have high moisture content. TCS foods (PHFs) require strict time/temperature control to prevent the growth of microorganisms.

Identifying Time/Temperature Control for Safety (Potentially Hazardous) Foods

Acidity

Foods with higher acidity, such as vinegars, tomatoes and citrus foods, inhibit the growth of microorganisms. Milk products, meats and vegetables, which tend to be neutral in pH, will support the growth of microorganisms if not handled correctly.

Water content

Foods with low water activity, such as breads, dried meats and jerkys, do not support the growth of microorganisms. Foods with high water activity, such as custards, meats and melons, support the growth of microorganisms.

Water activity is indicated by the symbol A_w. An A_w of 0.85 or lower inhibits growth while an A_w of 0.85 or above supports the growth of microorganisms.

Preventing Time/Temperature Abuse

As you learned in the section on biological contamination, the **temperature danger zone** is any temperature from 41°F to 135°F (5°C to 57°C). Within that range, bacteria grow even faster from 70°F to 125°F (21°C to 52°C).

In order to protect food, you need to minimize the amount of time food spends in the temperature danger zone.

Time/Temperature Controlled for Safety Foods

Milk and milk products

Meat

Melons

Green leafy vegetables

Insert the thermometer completely into the liquid without touching the sides or bottom of the container.

Example of cross-contamination

The following procedures will help you reduce the chance of time/temperature abuse:

- Ensure that the right thermometers and timers are readily available. Ideally, each employee should have their own thermometer. Thermometers with glass stems must not be used.

- Determine which foods need to be monitored and when. Assign specific staff members the responsibility of monitoring these foods and require them to keep a record of the foods they have monitored.

- Incorporate time/temperature controls into your daily practices. For example: require that staff remove from the refrigerator only amounts of food that can be prepared in a timely manner, use cold food prep items when preparing salads that contain TCS foods (PHFs) and always cook TCS foods (PHFs) to the required minimum internal temperatures.

- Set policies that make clear to employees the steps to be taken if food is time/temperature abused.

Preventing Cross-Contamination

Cross-contamination is one of the most common—and dangerous—types of bacterial contamination. **Cross-contamination** occurs when bacteria from contaminated foods—usually raw—transfers to other foods. Cross-contamination can happen in a variety of ways:

- Direct—for example, when raw meat touches another food

- By drip—for example, when raw meat incorrectly stored above ready-to-eat food drips blood onto that food

- Or indirect—for example, when raw food and ready-to-eat food are prepared with the same equipment

Since food is susceptible to all methods of cross-contamination at all stages from purchase to service, it's vital that you have controls in place to prevent cross-contamination.

Whenever possible, use different prep areas to prepare poultry, raw meat and fish than are used to prepare ready-to-eat foods. If separate prep areas are not possible, prepare foods at different times and be sure to clean and sanitize the area before moving from raw meat, poultry and fish preparation to ready-to-eat food preparation.

Properly store food. For example, raw food such as eggs, meat and poultry must always be placed below TCS (PHF) and ready-to-eat foods, such as lettuce and tomatoes. Frozen, commercially processed and packaged raw animal food may be stored or displayed with or above frozen, commercially processed and packaged, ready-to-eat food.

Never use the same equipment for handling raw and ready-to-eat food without it being thoroughly cleaned and sanitized in between those food types. Use a color coding system to reduce the risk of cross-contamination by ensuring the same equipment is not used for different food types. Color coding does not eliminate the need to clean and sanitize equipment. Color coding can be used for many types of equipment such as cutting boards, knife handles, work surfaces, cloths, protective clothing and packaging material. Systems of color coding may differ from one establishment to another, so it is important that your staff know and understand the color code in your workplace.

PURCHASE AND DELIVERY

LESSON 2

After completing this lesson, you should be able to:

- Identify the safety hazards associated with the purchase and delivery of food
- Implement controls to help avoid the potential hazards
- Use monitoring actions and set delivery standards to ensure the efficiency of the controls

Recognizing Potential Hazards

There are two potential, significant safety hazards you will face when dealing with the purchase and delivery of food. One potential hazard is the contamination of food by physical contaminants, chemicals or bacteria, either before or at the point of delivery. Another hazard is the multiplication of bacteria within the food, either "en route" to the establishment or at the point of delivery.

Avoiding Hazards

Choose the safest ingredients

Some food items, such as frozen raw chicken and raw shell eggs, may present an unacceptable risk. This risk can often be reduced by choosing alternative ingredients, for example, fresh or cooked chicken and pasteurized eggs. Fish and shellfish are required to be commercially and legally caught and harvested. Shellfish must have certification because of the possibility of naturally occurring environmental contaminations. Shellfish must arrive with identification tags, and the tags must be kept on file for 90 days. This allows you to trace any shellfish responsible for a foodborne illness back to the date of sale. For this reason, it is vital that you not mix batches of shellfish.

Use reputable suppliers

A reputable supplier is one with an outstanding record in food safety. This is demonstrated by the food safety management system, which should include procedures to ensure food safety or a HACCP program (described in Chapter 9). Suppliers should also provide food safety training to their employees. Reputable suppliers ensure the condition of food delivered and the standard of the driver and vehicle. A reputable supplier gets its food products from approved sources that meet all local, state and federal regulations and operate under government inspection. This is especially important with regard to produce, dairy, egg and seafood suppliers. For example, sealed foods, such as low-acidity canned foods and fluid milk, must be purchased from a processing plant governed by a food regulatory agency such as the FDA or the USDA. Finally, a supplier cannot offer food prepared in a private residence.

Set expectations with your supplier

By specifying to the supplier exactly what you expect with regard to the standard of food items and delivery times, you reduce the risk of receiving unsuitable or unfit food. For example, you should request that deliveries be made when your staff is free to inspect the deliveries. Employees must verify that deliveries of food during nonoperating hours are from approved sources and are placed into appropriate storage locations to ensure that they are maintained at the required temperatures, protected from contamination, and unadulterated, pure and safe for consumption.

Ensure that refrigerated trucks are used for delivery when necessary to maintain proper temperature control.

Check a clock for accuracy

Choose food items that are packaged appropriately

Food items without appropriate packaging are at greater risk of contamination. Delivered food items should never be left outside for prolonged periods prior to storage.

Unpack packaged food items in separate designated areas

Having a separate unpacking area will ensure that any contaminants on the surface of a food item can be safely spotted and removed, thus preventing its introduction to the food premises and other food items.

Maintain proper temperature control

Proper temperature control must be maintained, even during delivery. For chilled or frozen goods, an appropriate refrigerated vehicle is usually required. For example, raw eggs must be delivered via equipment that maintains an ambient temperature at or below 45°F (7°C). Refrigerated foods must be received at 41°F (5°C) or colder; 38°F (1°C to 3°C). You can even receive meats, poultry and seafood as low as 28°F (−2°C) without risk of damage from freezing. Fresh fish should optimally arrive alive and be packed on ice and have an internal temperature between 32°F and 41°F (0°C and 4°C). Check your local regulations for specified receiving temperature other than 41°F (5°C). Exceptions may apply for live molluscan shellfish and/or shell eggs. Frozen food should be received below 0°F (−18°C). Reject foods that show signs of thawing and refreezing or open containers. Thawing and refreezing can be indicated by clumping, ice crystals in the food and puddles from frozen water at the bottom of a container.

Store food immediately after delivery

Following delivery, all food items must be thoroughly checked, outer packaging removed, and stored correctly to prevent bacterial multiplication. Time/temperature control for safety (TCS) foods are at a greater risk from bacterial multiplication and should be stored quickly, within 15 minutes of delivery.

Audit your suppliers' premises, records and systems

Reputable suppliers will have impeccable premises, with full records of the products going in and out. They should also have food safety management systems in place, and, if applicable, you may want to check their HACCP documentation to ensure that their approach to food safety mirrors yours. You need to do your own research on your supplier. Ask other people in your field about their experiences with specific suppliers. If it turns out that they have had bad experiences, for example, damaged packaging or spoiled food, you should look elsewhere. Whenever possible, you should inspect your suppliers' warehouses or plants to ensure that they are clean and well run.

Check that the temperature of the delivered food items is acceptable

Deliveries of chilled food above 41°F (5°C) and frozen food above 0°F (−18°C) must be rejected. Whoever checks the delivery should ensure that the shortcoming is communicated to the delivery driver and stated on the delivery note. That person must also inform you immediately, as the incident will require that you investigate further with the supplier. It may even result in a change of supplier.

Check the condition of the delivery vehicle and its driver

The delivery vehicle must be hygienic and must be refrigerated if transporting chilled or frozen goods. There should also not be any materials on the vehicle, such as toxic chemicals or machinery, that could contaminate the food delivered. The driver should have impeccable personal hygiene and should be capable of delivering goods correctly, for example, avoiding the cross-contamination of foods within a large mixed delivery. If your monitoring checks spot deficiencies, then the reputability of your supplier may be put into question.

Check the condition of packaged food items

Boxes and packaging must be clean and free from any signs of pest infestation, contamination, dampness or damage. If any of these are spotted, you need to reject the product. You also need to check canned and vacuum-packed foods closely because *Clostridium botulinum* can grow in foods packaged this way. Reject food if the packages are swollen or show signs of leaking. If canned food is rusted or dented, or if one end of a can pops when the other end is pushed, it must be rejected.

Check canned and vacuum-packed foods closely because *Clostridium botulinum* can grow in foods packaged this way.

Check all products' date labels

Delivered food items must be checked to ensure they are not out-of-date. Additionally, care must be taken to ensure they will not expire too soon for their intended use. Any products that are not labeled must be rejected. Shortcomings in this respect are rather telling of problems with a supplier, and you should investigate further.

Check that the quality and quantity of the delivered items match those ordered and listed on the delivery note

Whoever checks a delivery needs to check the delivery note or purchase order and ensure that the quantities delivered are as per the note or purchase order. The same checks are needed regarding the type or quality standards of the food delivered. If any food does not meet the standards you set, it should be rejected. When rejecting food, make sure that your staff keeps the rejected product away from other food, records the rejection on the delivery note, receipt or purchase order, informs the delivery person of the rejected product and gets a signed adjustment or credit slip from the delivery person before the rejected product is removed from the establishment. In addition, any shortcomings should be reported to you, and you should investigate the matter further with your supplier. Check delivery monitoring records regularly.

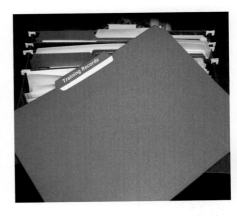

STORAGE

LESSON 3

After completing this lesson, you should be able to:

- List the order in which received goods should be stored
- Describe the controls and monitoring procedures for each type of delivered good
- Identify the guidelines for storing specific food items

> ### >TERMS TO KNOW
>
> **STOCK ROTATION** The practice of ensuring that the oldest stock is used first and that all stock is used within its shelf life.
>
> **FIFO** The acronym for first in first out, which is used in stock rotation.

The three main types of delivered goods received by food establishments are refrigerated items, frozen items and dry goods. It is important to follow the proper guidelines for storing these items to prevent spoilage and contamination.

Stock Rotation

It is essential that newly delivered food items be stored below or behind older stock. In other words, stored goods must be correctly rotated. Correct **stock rotation** ensures that the food used is safe and of good, consistent quality by avoiding spoilage, reducing the risk of pest infestation and reducing the risk of food going out-of-date. Remember the rule—"first in, first out," or **FIFO.** If your staff knows and understands this, then older food will always be used first and you will avoid waste and save money.

Stock Records

Written stock control records are also recommended. Written records will help maintain correct stock levels and determine when out-of-date food should be discarded. It helps to identify any necessary corrective actions, such as further training for your staff on date labels or stock rotation principles. All TCS foods (PHFs) and perishable foods should be checked daily.

Out-of-date foods must be removed and destroyed or stored separately pending appropriate action. You should require employees to notify you in all cases.

Labeling

Food labels must indicate the food name and the use-by date.

Whenever possible, food should be stored in its original packaging. If you transfer food to a new container, the container must be clean and sanitized. The container must also be labeled with the food name and the original use-by date. Labeling foods is especially important because certain foods can be difficult to identify if not in the original packaging. This can lead to allergic reactions when customers consume food items to which they are allergic. It can even result in accidental chemical poisoning.

Refrigerated Storage

The major hazards of refrigerated food storage are the contamination of foods by raw foods and the multiplication of bacteria or spoilage organisms when temperatures are too high or storage is prolonged. There are controls that can be put into place to avoid the hazards associated with refrigerated storage.

For starters:

- Nearly all TCS foods (PHFs) must be stored at or below 41°F (5°C). Some foods, such as fresh fish, should be stored at lower temperatures. The maximum shelf life for any TCS food (PHF) and ready-to-eat food prepared in your establishment is seven days. Any TCS foods (PHFs) or ready-to-eat foods stored longer than seven days or at an incorrect temperature must be discarded. In order to maintain this operating temperature, refrigerators must be located in well-ventilated areas, away from heat sources such as ovens.

- Refrigerators should have open shelving, and the refrigerator must not be overloaded. Congested shelving prevents cold air from circulating and can raise the temperature of the refrigerator and food stored in it. You must also not use refrigerators for cooling food.

- Alarmed units are recommended, so that you are warned of unacceptable temperatures. Alarms may also be installed so that you are automatically informed of high humidity values. High humidity will encourage the growth of spoilage bacteria.

- Keeping refrigerator doors closed as much as possible will help maintain proper temperatures. Install cold curtains on frequently used walk-in refrigerators.
- Ideally, ready-to-eat foods and raw foods should be kept in separate refrigerators. If in the same unit, raw food such as eggs, meat and poultry must always be placed below TCS food (PHF) and ready-to-eat food, such as lettuce and tomatoes, to avoid cross-contamination.
- Stock should be rotated so that older food is always to the front of the refrigerator and gets used first.
- Refrigerators must be cleaned and sanitized regularly, to prevent the accumulation of dirt and to remove any physical contaminants. Care must be taken to prevent chemical contamination of food items or refrigerator surfaces while cleaning and sanitizing.
- Food should be properly covered to prevent drying out, cross-contamination and absorption of odor. For example, storing uncovered onion slices in a refrigerator will cause other foods to take on the onion odor. Simply covering an open can and placing it in the refrigerator is not acceptable. You should first empty the unused contents of the can into a suitable covered storage container and apply the proper labeling.

Check refrigerator temperature regularly throughout the day

All refrigerators should be fitted with accurate thermometers with liquid crystal displays. The relationship between the displayed temperatures and the temperature of the food in the warmest part of the refrigerator—normally the top—should be determined. The display temperature should be checked every time the refrigerator is used. The temperature of food, or a food substitute, should be checked with an accurate, calibrated digital thermometer at the start of the day, at the end of the day and whenever the display reading is unacceptable. A record of food temperatures should be maintained and checked weekly. Sophisticated, automatic monitoring and alarm systems are becoming more common, along with data loggers.

Audit refrigerators on a regular basis

Regular audits of refrigerators should address excessive temperatures and humidity, as well as how staff loads and cleans the refrigerator. Audits should also involve checking for any electrical problems, refrigerant leaks and damage to door seals, as they may all contribute to an increase of the refrigerator temperature. A maintenance contract is recommended to ensure that refrigerators are kept in good repair.

Check the condition of food daily

Unfit or contaminated food should be discarded immediately. All food in refrigerators should be in appropriate covered containers. If any open cans are spotted they should be rejected, as should any food in damaged containers. Food must never be stored in empty chemical containers. The base of storage containers should not come into contact with open food.

Check date labels

Out-of-date food must be rejected at once. All use-by date labels should be checked daily and stock rotated as appropriate.

Frozen Storage

The major hazard associated with frozen food storage is the multiplication of dormant bacteria if the temperature rises above 14°F (–10°C). Unfortunately, freezing does not kill bacteria. Frozen food should never be left out at

room temperature. As soon frozen food has been delivered and inspected, it should immediately be placed in the freezer. Freezer temperature should be maintained at 0°F (−18°C). Do not refreeze food.

In order to maintain the correct operating temperature, freezers must be located in well-ventilated areas away from heat sources, and food must not be stored above the freezer load line. Keeping freezers closed as much as possible will help maintain proper temperatures. Other things to keep in mind for frozen storage: install cold curtains on freezers that are frequently used; do not store warm food, as this can result in other foods thawing; before freezing any food cooked on site, label the package with the content and its use-by date and observe the principle of proper stock rotation—that is, older food must always be kept above or in front of longer-life food. Raw meat, poultry, and seafood can be stored with or above RTE food in a freezer if all of the items have been commercially processed and packaged.

It is essential to follow the manufacturer's instructions, as the quality of food gradually deteriorates, even in the best of freezers. For example, vegetables, fruit, and most meat can be frozen for up to 12 months, whereas pork, sausages, fatty fish, butter or soft cheese should only be stored for up to 6 months.

Check freezer temperatures regularly, on a daily basis

All freezers should be fitted with accurate thermometers, and temperatures should be checked regularly, at least daily. A record of temperatures should be maintained, and you should check the record regularly.

Audit freezers on a regular basis

Regular audits of freezers should address temperatures, food load and cleaning. Audits should also involve checking for any electrical problems, refrigerant leaks and damage to door seals. A maintenance contract is recommended to ensure freezers are kept in good repair.

Defrost freezers on a regular basis

Freezers are most efficient if they are frost-free. Before defrosting a freezer, first move its contents to another freezer.

Check the condition of food on a daily basis

Unfit or contaminated food should be discarded immediately. All food in freezers should be in appropriate, covered containers.

Check date labels

All "best before" date labels should be checked weekly, and stock should be rotated as appropriate. Any out-of-date food should be discarded.

Dry Storage

The hazards associated with dry food storage are physical contamination—for example, by objects brought in on delivery packaging, such as staples or cardboard—or chemical contamination from rusty cans or chemicals being kept in the dry storage area. Multiplication of food poisoning bacteria or food spoilage microorganisms is also a hazard if the food becomes damp.

Dry storage areas need to be well maintained and kept clean. The areas should be kept dry and maintain a temperature between 50°F and 70°F (10°C and 21°C). They also must be well-lit and ventilated.

Shelves must be easy-to-clean. Slotted shelving units improve air circulation. Do not store food in hallways, utility rooms, restrooms or sleeping and dressing rooms.

Food should be stored at least six inches off the floor and away from walls to enable thorough cleaning and maintenance work, protect it from condensation and contamination and easily check for signs of pest infestation.

Regular inspections will ensure that dry storage areas are pest-free. Correct stock rotation will ensure that older food is used first but will also ensure that any problems with food pests are spotted.

Covering food prevents contamination from pests, foreign objects or leaking food from damaged packaging. Additionally, storing vegetables heavily contaminated with soil on the bottom rack will protect other food items in the storage area. Potatoes should be stored in the dark to prevent them from sprouting or turning green.

Keeping food and cleaning chemicals in separate storage areas will prevent contamination and the tainting of food by the chemicals.

Audit dry storage areas on a regular basis

Regular audits of dry storage areas should address temperature, lighting and ventilation as well as how staff loads and cleans the storeroom. In particular, you should check that food is stored correctly, on the appropriate shelves and following good stock rotation. Food should be stored away from walls or in mobile, rodent-proof bins, and there should never be any food left on the floor.

Check for signs of pest infestation

You should check less accessible areas for cleanliness and any signs of pest infestation, as well as the condition of the food and food packaging itself.

Check the condition of food

Foods should be kept in their original packaging or airtight containers and rotated correctly. Any spoiled food or food that is out-of-date should be disposed of. There should never be any open cans or bottles stored in the dry storage area.

Check the condition of food packaging

Check that food packaging is in good condition, showing no signs of dampness, damage or pest infestation. You should also check that cans are not blown, badly dented, holed, rusty or damaged at the seams. If any defects are found, the cans must be thrown away. Before using canned food, wipe the top with a clean, sanitized cloth to prevent dirt from falling into the can.

Check date labels

Out-of-date food must be discarded at once. All "best before" date labels should be checked at least weekly and stock rotated as appropriate. Be sure to check stock control records weekly.

Storing Specific Foods

While there are several storage principles that apply to all food, some specific storage principles apply to certain foods.

Meats

Meats, such as beef, pork, lamb, veal and wild game, need to be stored at 41°F (5°C) or below and must be USDA inspected. If meat is delivered frozen, it should be stored in the freezer in a timely manner so that it does not have an opportunity to thaw. Meat should be stored in either its original packaging, in clean and sanitized containers or in airtight, moisture-proof wrapping. Ground or otherwise non-intact meats must be stored below whole-muscle intact cuts of meat unless they are packaged in a manner that prevents the potential for cross-contamination. Meats have a short shelf life and if time/temperature abused, support rapid bacterial growth. Before storing and preparing meat, ensure that it is firm and elastic and has a fresh odor. Discard any meat that appears to be spoiled.

Foods with specific storage principles include:

- Meats
- Eggs
- Poultry
- Seafood
- Milk and dairy
- Fresh fruits and vegetables
- Reduced oxygen packaging (ROP) foods

Eggs

Eggs should be stored at 45°F (7°C) or lower. Eggs should not be subjected to fluctuating humidity or temperatures, which encourage condensation. If you are using a reputable supplier, eggs will have been washed and sanitized before delivery. Therefore, don't wash eggs before storing them. If egg products are delivered frozen, they should be stored in the freezer in a timely manner so that they do not have an opportunity to thaw. Nonfrozen, liquid eggs need to be stored according to the manufacturer's recommendations. Dry egg products should be stored under dry storage controls and monitoring. After dry eggs are mixed with water, they need to be stored at 41°F (5°C) or lower. Whether you are using fresh eggs or other egg products, it's important that you work with small amounts so that they are not left out at room temperature.

Poultry

Poultry includes chicken, turkey, goose and duck and should be stored at 41°F (5°C) or below. Poultry must also be USDA inspected. If poultry is delivered frozen, it should be stored in the freezer in a timely manner so that it does not have an opportunity to thaw. If poultry is delivered on ice, it can be stored in a refrigerator as is, provided that it is stored in a self-draining container that is cleaned and sanitized regularly. Also, you must change the ice often. Poultry should be stored in either its original packaging, in clean and sanitized containers or in airtight, moisture-proof wrapping. Poultry has a short shelf life and if time/temperature abused, supports rapid bacterial growth. Before storing and preparing poultry, ensure that the flesh is firm and elastic. Poultry can be a source of Salmonella and Campylobacter.

Seafood

Seafood includes fish, shellfish and crustaceans. Fresh fish should be stored at an internal temperature between 32°F and 41°F (0°C and 5°C). If fish is delivered frozen, it should be stored in the freezer in a timely manner so that it does not have an opportunity to thaw. Fish fillets and steaks should be stored in their original packaging or securely wrapped in moisture-proof wrapping. If fresh, whole fish is delivered on ice, it can be stored in a refrigerator as is, provided that it is stored in a self-draining container that is cleaned and sanitized regularly. Also, you must change the ice often. Fresh fish should have a firm, elastic flesh, clear eyes and bright red gills. Fresh fish must not have cloudy eyes or a fishy odor. Live shellfish should be stored at a temperature of 45°F (7°C) or below. If you want to display live molluscan shellfish in a tank, they need to be for display only. If you wish to serve them, you need to contact your local regula-

Smoked fish should be kept in the refrigerator below 36°F (2°C) and consumed within 14 days after smoking. For longer storage, the fish may be frozen immediately after smoking. Store smoked fish in the freezer for no longer than two to three months.

tory agency to get a variance. Smoked fish should be kept in the refrigerator below 36°F (2°C) and consumed within 14 days after smoking. For longer storage, the fish may be frozen immediately after smoking. Store smoked fish in the freezer for no longer than two to three months.

Milk and dairy

Milk and dairy products, including bakery fillings containing dairy, must be stored at 41°F (5°C) or lower. The use-by or expiration date on milk and dairy products represents the last day you can sell or use the product. Fluid milk must be Grade A pasteurized, and butters and cheeses must be free of contamination. Dairy containers cannot be reused or refilled.

Fresh fruits and vegetables

Many whole, raw fruits and vegetables, such as carrots and celery, can be refrigerated at 41°F (5°C) or below, at a relative humidity of 85 to 95 percent. Other whole fruits and vegetables, such as citrus fruits, root vegetables and hard-rind squash can be stored in dry storage. The ideal temperature for storing fruits and vegetables in dry storage is 60°F to 70°F (16°C to 21°C). As you plan for food preparation, keep in mind that many fruits and vegetable continue to ripen after they are harvested. Fruits and vegetables, such as avocados, bananas and tomatoes, ripen best at room temperature. Because moisture promotes bacterial growth most produce should be washed before preparation or service, not upon delivery. Prior to washing, fruits and vegetables must be stored away from ready-to-eat foods. Never store or soak multiple foods or mixed batches of the same food item.

ROP foods

Reduced oxygen packaging (ROP) encompasses a large variety of packaging methods where the internal environment of the package contains less than the normal ambient oxygen level. Using ROP methods in food establishments has the advantage of providing extended shelf life to many foods because it inhibits spoilage organisms that are typically aerobic. ROP foods should be stored according to the manufacturer's recommendations or at 41°F (5°C) or below. If ROP food is delivered frozen, it should be stored in the freezer in a timely manner so that it does not have an opportunity to thaw. While vacuum packaging will slow the growth of microorganisms that require oxygen to grow, it will not prevent microorganisms that do not require oxygen to grow. Because ROP foods are susceptible to *Clostridium botulinum* growth, discard the product if the package shows signs of contamination, such as slimy packaging or if it contains an excessive amount of liquid. While labeling is important on all packaged foods, it's especially important the ROP foods have labels that list the use-by date, contents, storage temperature and preparation instructions.

FOOD PREPARATION

LESSON 4

After completing this lesson, you should be able to:

- Identify the hazards, controls, and monitoring actions involved in food preparation

TERMS TO KNOW

SLACKING The process of gradually increasing frozen food from a temperature of −10°F to 25°F (−23°C to −4°C) to facilitate even heat distribution during the cooking process.

General Preparation

The main hazards likely to occur during food preparation are cross-contamination and time/temperature abuse that result in the multiplication of bacteria.

Prepare raw foods and ready-to-eat foods in separate areas

Ready-to-eat foods are at risk of cross-contamination from raw food if the same areas and equipment are used without being thoroughly cleaned and sanitized. Ideally, separate areas and equipment should be used. The color coding of equipment will help ensure equipment is not misused.

Have high standards of personal hygiene

Food handlers are a major source of bacterial and physical contamination, especially if they have poor personal hygiene. High standards of hygiene include: washing hands properly and regularly, keeping hair clean and covered, wearing clean, protective clothing, covering cuts with waterproof bandages or finger cots, and keeping nails clean, rounded, and short. Food handlers experiencing diarrhea or vomiting, a bad cold, boils, or septic cuts must report their symptoms to their supervisor and should not handle food.

Minimize the handling of food

Hands are a major vehicle of contamination, especially if they are not kept clean. There should be separate sinks for washing raw food and sinks for hand washing. Ideally, food handlers dealing with raw food and those dealing with TCS foods (PHFs) should also wash their hands in separate basins to avoid cross-contamination. Once hands are clean, food handlers should wear gloves to inhibit the spread of bacteria. Suitable equipment, such as clean tongs, forks, and cake slicers should also be used to minimize hand contact with food.

Use disposable towels

Wiping hands on a cloth towel will only transfer the bacteria from hands to the cloth, which then acts as a vehicle to spread bacteria around the food preparation room. To avoid contamination, food handlers should use disposable towels. The waste container should be pedal-operated, so as to avoid cross-contamination via the container's lid.

Minimize the amount of food prepared

Using batch cooking limits the amount of food that is exposed to hazards. If too much food is prepared there is a risk of bacterial multiplication if it is not stored correctly. There may be a reduction in quality if there is a long gap between preparing and using or serving food. Always make sure food is not prepared too far in advance, as this is one of the most common causes of food poisoning.

Minimize the time in the temperature danger zone

Most foodborne illness causing bacteria multiply in the temperature danger zone, between 41°F and 135°F (5°C and 57°C). Room temperature is well within this temperature range. So, once food is prepared, it should immediately be used or refrigerated. By occasionally checking the temperature of food during preparation, you will be able to determine a maximum safe time for preparation. If you are putting food at risk of bacterial multiplication, remedial action such as changes in preparation procedures or recipes should be introduced.

Follow a "clean as you go" policy

Food handlers should work in a logical manner, ensuring work surfaces are kept as tidy as possible and effectively clean and sanitize areas and equipment used for preparing food. Food spills and waste should be cleaned up immediately.

Audits

Regular audits of preparation procedures and activity will allow you to check whether your staff is carrying out its duties safely, and as per your instructions. It will also give you an indication as to whether any further training on personal hygiene is needed, or whether you could introduce any improvements to the workflow, to minimize the risk of cross-contamination.

Thawing

Potential hazards in the thawing of frozen, raw food items are cross-contamination from thawed liquid to ready-to-eat food and the multiplication of bacteria that were dormant while the food was frozen.

The refrigerator method

Thaw food in a refrigerator that maintains the food temperature at 41°F (5°C) or colder.

The submergence method

Completely submerge the frozen food item under clean, cold running water at 70°F (21°C) or below for a period of time that prevents the temperature of any thawed portion from rising above 41°F (5°C).

The water flow must be strong enough to allow particles to float off into an overflow drain.

When thawing with water, you must clean and sanitize the thawing area before and after thawing food.

The cooking process

Food can be thawed as part of the cooking process, provided that it reaches the minimum internal cooking temperature.

Slacking may be used to gradually thaw frozen food, such as shrimp and breaded chicken breasts, in preparation for deep-fat frying or to allow for even heating during the cooking process. Slacking is the process of gradually increasing the temperature of frozen food from −10°F to 25°F (−23°C to −4°C).

The microwave method

Thawing food in a microwave oven is acceptable, provided that it will immediately be cooked afterward.

Food must not be thawed at room temperature. If thawed food is left at room temperature, then any dormant bacteria will start to multiply within the food.

To help control the hazards associated with thawing, thaw raw food in an area entirely separate from ready-to-eat foods. Thawing raw food in an area separate from ready-to-eat foods helps to avoid the cross-contamination of ready-to-eat foods from thawing, raw food. In particular, raw frozen chickens must never be thawed in the same area as cooked food items cooling before refrigeration. The areas used for thawing should be thoroughly cleaned and sanitized after the food is removed. Another control is to refrigerate or cook food immediately after thawing.

Monitoring the Thawing of Food

The main monitoring action needed regarding thawing is to audit your thawing processes on a regular basis. Regular audits should allow you to check whether better and safer thawing processes can be introduced. They will also allow you to ascertain if your staff needs further training to ensure that they know how to check if thawing is complete.

COOKING

LESSON 5

After completing this lesson, you should be able to:

- Describe the hazards involved in the cooking process
- List the controls and monitoring actions that can help avoid the hazards
- List the guidelines for cooking specific types of foods

The main hazards in the cooking, or processing, stage are the survival of bacteria as a result of inadequate cooking, the multiplication of bacteria as a result of prolonged cooking at low temperatures and contamination. Contamination can be caused by various sources; for example, repeated tasting with the same unwashed spoon, insects falling into uncovered cooking pans and the use of equipment made out of inappropriate materials, such as copper or aluminum, for cooking acidic foods.

Cook food thoroughly and handle it properly

You must use a thermometer to ensure that food is cooked thoroughly. However, if the thermometer is not cleaned and sanitized, it may act as a vehicle of contamination. As spores and some toxins can even survive boiling, it is important to note that cooking without using a thermometer to check the food's internal temperature cannot be relied on to guarantee safe food. Food must always be cooked to the minimum internal temperature for the required amount of time.

Ensure that pans and utensils are made of suitable materials, clean and free of physical contaminants

Pans and utensils will be vehicles of contamination if they are not thoroughly cleaned between uses. Additionally, if copper or aluminum equipment is used for cooking acidic food, chemical contamination of the food may result.

Ensure that the heat source is under the whole pan base

If the base of a pan being used for cooking is bigger than the heat source, or if stirring is not frequent, then there will not be an even distribution of heat throughout the food. Therefore, there will be parts of the pan where bacteria will not be destroyed; instead, they may multiply to large numbers.

Cover pans when not stirring

Foreign objects may fall into uncovered pans. Flying insects, either dead or alive, are a potential hazard.

Taste food only with a clean, sanitized spoon

Tasting food with fingers or with dirty spoons may result in bacterial or physical contamination of food.

When cooking in a microwave take the appropriate precautions

- Rotate or stir the food during the cooking process to ensure even heat distribution.
- Ensure that all parts of the food reach a temperature of at least 165°F (74°C).
- Cover the food to help maintain moisture.
- Let food sit for two minutes to ensure temperature stability.

There are monitoring actions that should be taken to ensure that your cooking controls are working effectively. Regular audits will allow you to determine where further training is needed for your staff on best working practices, personal hygiene, or the use of thermometers. They will also allow you to check whether better and safer cooking processes can be introduced. Regularly checking cooking times and internal temperatures of food will allow you to confirm that your staff is following the correct procedures and instructions. Ensure that you check cooking temperature monitoring records regularly.

Holding food guidelines

Once food is prepared, it must be held at the proper temperature until it is served to prevent microorganisms from multiplying and making a customer ill. Hot food must be held at 135°F (57°C) or higher. Cold food must be held at 41°F (5°C) or lower.

Cooking Specific Foods

Cooking food to its required minimum internal temperature is the only way to reduce the amount of bacteria in food items. The minimum internal temperature is not the same for all food items. Minimum internal temperature must be reached and held for a certain amount of time.

The information presented below covers the minimum internal cooking temperatures for commonly consumed food items.

Eggs	Ground Beef, Pork, other Meats	Beef Steaks	Poultry	Fish	Pork, Veal, Lamb Chops
Raw eggs cooked to order for immediate service must be cooked to at least 145°F (63°C) for 15 seconds. Eggs that will be held for service later must be cooked to 155°F (68°C) for 15 seconds.	Because grinding can spread contaminants throughout the meat, ground meats must be cooked at 155°F (68°C) for 15 seconds.	Steaks must be cooked to at least 145°F (63°C) for 15 seconds.	Poultry must be cooked to at least 165°F (74°C) for 15 seconds.	Fish must be cooked to at least 145°F (63°C) for 15 seconds.	Pork, veal and lamb chops and tenderloin medallions must be cooked to at least 145°F (63°C) for 15 seconds.

Tenderizer-Injected and Mechanically Tenderized Meats	Stuffing and Stuffed Foods	Fruits and Vegetables	Commercially Raised Game and Birds	Leftovers	Ready-to-Eat Foods
Beef and pork injected with tenderizers must be cooked to at least 155°F (68°C) for 15 seconds.	Stuffing made with TCS foods (PHF), stuffed fish, stuffed meat, stuffed poultry and stuffed pasta must be cooked to at least 165°F (74°C) for 15 seconds.	Fruits and vegetables being cooked for hot holding should be cooked to at least 135°F (57°C).	Farm-raised game, including elk, deer and bison can follow the same minimum internal temperature guidelines as beef. Steaks should be cooked to 145°F (63°C) for 15 seconds; ground meat to 155°F (68°C) for 15 seconds; stuffed meats to at least 165°F (74°C) for 15 seconds.	Leftovers, or previously cooked TCS foods (PHF), must be cooked to at least 165°F (74°C) for 15 seconds.	Ready-to-eat food must be commercially processed in airtight containers or intact packaging from a government-regulated food processing plant. It must be heated to at least 135°F (57°C) within 2 hours.

Due to the potential of bacterial illness in a highly susceptible population, the FDA Food Code prohibits offering raw or undercooked comminuted (ground, chopped, minced; a mixture of two or more types of meat) meat on a children's menu.

Roasts

Roasts, including beef roasts, corned beef, pork roasts and ham, must be cooked to the following internal temperatures:

Temperature	Time (in minutes)
130°F (54°C)	112
131°F (55°C)	89
133°F (56°C)	56
135°F (57°C)	36
136°F (58°C)	28
138°F (59°C)	18
140°F (60°C)	12
142°F (61°C)	8
144°F (62°C)	5
145°F (63°C)	4

Beef or fish (cubed)

Beef or fish cut up into small pieces should be cooked according to the minimum temperatures:

Temperature	Time
145°F (63°C)	3 minutes
150°F (66°C)	1 minute
155°F (68°C)	15 seconds
158°F (70°C)	< 1 second

Noncontinuous cooking

Raw meat, poultry, seafood or eggs may be partially cooked during preparation and finished cooking just prior to service if:

1. The food is not cooked initially longer than 60 minutes.
2. The food is immediately cooled after heating and stored in a refrigerator or freezer following TCS food (PHF) requirements.
3. Prior to sale or service, all parts of the food are reheated to at least 165°F (74°C) for 15 seconds.
4. If neither hot held nor served immediately, the food must be cooled again following TCS food (PHF) requirements.

These procedures must be written and approved by the local regulatory authority. Documentation should include monitoring procedures and corrective actions, labeling of non-continuous foods and assurance of separation from ready-to-eat foods.

COOLING AND REHEATING

LESSON 6

After completing this lesson, you should be able to:

- Describe the potential hazards involved in the processes of cooling and reheating food
- List the controls that can be implemented to avoid these hazards
- Specify the monitoring actions to ensure that the controls are working effectively

>TERMS TO KNOW

HOT HOLDING The storage of cooked food at 135°F (57°C) or higher, while awaiting consumption by customers.

REHEATING The process of re-cooking previously cooked and cooled foods to a temperature of at least 165°F (74°C).

The processes of cooling and reheating food before serving it to customers are very important to the overall flow of food from purchase to service.

Cooling

Once cooked, food may require storing before serving. But before being stored under refrigeration, the food must first be allowed to cool down safely. The hazards associated with cooling include the multiplication of food poisoning bacteria not destroyed during cooking, as a result of either inadequate cooking or the activation of spores, and the contamination of food by bacteria, foreign bodies or chemicals while cooling.

Cool food quickly after cooking and either refrigerate it or use it as soon as it has cooled

Cooling food quickly after cooking and refrigerating or using it as soon as it has cooled will minimize the period of time the food is in the temperature danger zone. As you plan for food preparation, keep in mind that cooling times depend on the size, thickness and weight of food. TCS foods (PHFs) must be cooled from 135°F to 70°F (57°C to 21°C) within two hours and from 135°F to 41°F (57°C to 5°C) within a total of six hours. Cooling from 135°F to 70°F (57°C to 21°C) is the most critical, because of the potential for rapid multiplication of bacteria. If food does not get cooled to 70°F (21°C) within two hours it must either be reheated to 165°F (74°C) and cooled again or discarded.

Do not cool hot food in a refrigerator

As tempting as it may be, don't cool hot food in a refrigerator. Doing so will raise the temperature of the other foods in the refrigerator and cause condensation.

Use shallow, stainless steel pans when cooling food

Shallow, stainless steel pans transfer heat faster than other containers. The pans should be no more than four inches deep and filled no more than two inches from the top.

Use the right cooling method

If you have large amounts of food to cool, a good first step is to divide it into smaller amounts. Ideally, the food should be stirred and placed in a blast chiller immediately. You can also cool food by placing it in an ice bath or steam-jacketed kettle, stirring it with ice paddles or adding ice or cold water as the last step in the food preparation.

Separate cooling food from raw food

Separate cooling food from raw food, especially raw frozen food that is thawing. If cooling food contacts raw food, it may be contaminated by bacteria.

Previously cooked and cooled TCS foods (PHFs) must be rapidly reheated to an internal temperature of 165°F (74°C) for 15 seconds within two hours.

Protect food during cooling

Uncovered food is always at risk of contamination from bacteria, foreign bodies, pests or tainting from chemicals. However, be aware that covering food may reduce the speed of cooling.

Store food that has been cooled to 70°F (21°C)

Once food has been cooled to at least 70°F (21°C), it can be stored in the refrigerator. Be sure that there is room for air to circulate around the food and check the food to guarantee that it is cooled to 41°F (5°C) within a total cooling time of six hours.

The monitoring actions needed regarding cooling include auditing your cooling procedures on a regular basis. These audits will ensure that your procedures are safe and that your staff follows them. If you use blast chillers, you should also check to make sure these are kept in good order and repair. It is vital to cool TCS foods (PHFs) as quickly as possible. Check that staff follows procedures to minimize cooling times and that food is always refrigerated within two hours of cooking.

Reheating

Food that has been cooled after cooking will eventually be served to customers, either hot or cold. If food is to be served hot, then it will require **reheating**. The reheating of food presents the same potential hazards as cooking—that is, contamination and the multiplication and survival of bacteria.

Reheat food to the proper temperature

Previously cooked and cooled TCS foods (PHFs) must be rapidly reheated to an internal temperature of 165°F (74°C) for 15 seconds within two hours. Food that does not reach 165°F (74°C) within two hours must be discarded. Food that is being reheated for immediate service may be served at any temperature as long as it was previously cooked and cooled properly. A clean and sanitized thermometer is the best method to ensure food has been reheated thoroughly.

Use the proper equipment to reheat food

Food must be reheated with proper food preparation equipment; hot-holding equipment must not be used to reheat food.

Use proper microwave cooking procedures

When using a microwave to reheat TCS foods (PHFs) for **hot holding**, all parts of the food need to reach an internal temperature of 165°F (74°C). While heating, the food must be rotated or stirred, covered and allowed to stand covered for two minutes after reheating.

The monitoring actions required for the reheating of food are basically the same as for cooking—that is, to regularly audit the reheating procedures used and check reheating temperatures using a sanitized thermometer. Regular audits will allow you to determine whether further training is needed for your staff on best working practices, personal hygiene or the use of thermometers. Audits will allow you to check whether better and safer reheating processes can be introduced. Regularly checking reheating times and internal temperatures of food will allow you to confirm that staff is following the correct procedures and instructions. Ensure you check temperature records regularly.

SERVICE

LESSON 7

After completing this lesson, you should be able to:

- Describe the potential hazards involved with the service of food
- List the controls that can be put in place to avoid these hazards
- Describe the monitoring actions to take to ensure that your controls are working effectively

Service is the final stage in the process that began way back when the food was purchased from the supplier. As serving safe food is your goal, presenting the food to your customers is the culmination of everything you have been working toward.

The hazards associated with the service of food include the multiplication of food poisoning bacteria because of prolonged periods at room temperature, and contamination from food handlers, equipment, utensils or even price tags in the case of retail premises. Food left open to self-service is also at risk of contamination from the customer.

Serve food quickly after cooking or removal from refrigeration

If food is served quickly after cooking, cooling or reheating, it will avoid the multiplication of bacteria, by keeping it out of the temperature danger zone for prolonged periods.

Protect food prepared for self-service by keeping it at the proper temperature

If food is to be displayed, possibly for self-service, then cold food must be kept below 41°F (5°C). Ensure that the cold-holding equipment keeps food at this temperature. With the exception of whole fruits and vegetables and cut, raw vegetables, food cannot be stored directly on ice. Hot food must be kept above 135°F (57°C). The exception to this rule is roast beef and pork. If they have been cooked to 130°F (54°C), they can be held at 130°F (54°C). There are times when food can be held without temperature controls, but it's vital that you check your local regulations before doing so.

Dispose of food after a predetermined amount of time

Set appropriate standards to ensure food is disposed of after a predetermined amount of time. Preparing food in small batches will help minimize the chance of time/temperature abuse and also result in less waste from discarding food not consumed in the predetermined amount of time.

Keep food covered or protected

If food is displayed, customers should not be able to handle open food. Food should be pre-wrapped, where feasible, and there should be sneeze guards and plenty of serving utensils to reduce the risk of contamination from customers. The handles of serving utensils should be longer than the display dishes, to prevent them dropping into food.

Separate raw food from ready-to-eat food

Raw food should be served with separate equipment and utensils from ready-to-eat food to prevent cross-contamination. Ideally, it should be served by different food handlers, using separate serving counters.

Don't re-serve food

Food that has been previously served to a customer must not be re-served to another customer. This includes garnishes, uneaten bread and unsealed condiments. Generally, only unopened, wrapped foods, such as wrapped crackers and condiment packets, can be re-served.

Handle ice properly

Ensure that your staff understand that ice is a food and should be handled carefully, especially when preparing and serving beverages. Because freezing water does not destroy bacteria or toxins, ice can contain those dangerous bacteria or toxins. Therefore, only drinking (potable) water should be used to make ice.

Thoroughly clean and sanitize all equipment and utensils used for service

All equipment and utensils that are likely to come into contact with food should be kept in good condition and effectively cleaned and sanitized to prevent cross-contamination and the multiplication of bacteria. Staff should also be trained to handle equipment and utensils properly, for example, holding flatware only by the handle and gripping drinking glasses only around the outside and not touching the rim or the inside edge.

Audit service procedures on a regular basis

Regular audits should address the personal hygiene of food handlers, as well as cleaning procedures, segregation of raw and ready-to-eat foods and temperature control.

If displaying food on hot-holding counters or refrigerated displays, you must also check that these units are kept in good working order and their temperatures are in the correct range so as to prevent multiplication of bacteria.

Check food condition and temperature

Checking food for quality, general condition and appearance will help ensure customer satisfaction.

Food temperatures should be checked frequently. Hot-and-cold-held foods need to be checked at least every four hours. Use thermometers to check food temperatures; don't rely on the holding equipment's temperature gauge. Hot food should be stirred regularly to ensure proper heat distribution.

If you sell prepacked food, you must also check date labels and ensure that stock is correctly rotated. These checks will help confirm that your service procedures are effective and safe.

Except when packaging food using a reduced-oxygen packaging method, ready-to-eat, TCS foods (PHFs) prepared and held in your food establishment more than 24 hours, must be clearly marked with the date by which the food must be eaten, sold, or discarded when held at 41°F (5°C) or less for a maximum of 7 days. The day of preparation should be counted as day one.

Freezing water does not destroy bacteria or toxins, therefore, only drinking (potable) water should be used to make ice.

Handle equipment such as flatware and drinking glasses appropriately to prevent contamination.

ASSESSMENT

1. All of the following are examples of TCS foods (PHFs) except:
   a. Milk, milk products and eggs
   b. Meat, poultry, fish, shellfish and crustaceans
   c. Sliced melons
   d. Sliced bread

2. What are the two most important things to do when protecting food?
   a. Prevent air exposure
   b. Prevent time/temperature abuse
   c. Prevent cross-contamination
   d. Prevent loss of moisture

3. What are the three methods of cross-contamination?
   a. Direct, drip, inverse
   b. Direct, indirect, drip
   c. Direct, indirect, inverse
   d. Direct, inverse, diverse

4. Which of the following is true regarding food safety and stock rotation?
   a. All TCS (PHF) and perishable foods should be checked daily
   b. All TCS (PHF) and perishable foods should be checked weekly
   c. All TCS (PHF) and perishable foods must be used within 14 days
   d. Food can be used or sold up to three days past the expiration date

5. Each of the following statements is true regarding the proper storage of meat except:
   a. Meat needs to be stored at 41°F (5°C) or below
   b. Before storing and preparing meat, ensure that it is firm and elastic and has a fresh odor
   c. Meat should be stored in its original packaging, or in clean and sanitized containers or airtight, moisture-proof wrapping
   d. Meats can be used or sold up to three days past their use-by date

6. Employees can help ensure food safety during preparation by:
   a. Minimizing the handling of food
   b. Using disposable cloths for wiping hands
   c. Following a "clean as you go" policy
   d. All of the above

7. Why is it so crucial to follow safe practices when thawing food?
   a. Because improper thawing can result in pest infestations
   b. Because dormant bacteria remain in frozen food and will start to multiply when the food's temperature is raised
   c. Because improper thawing can cause freezer burn
   d. Because the thawed liquid can contaminate ready-to-eat food

8. All but which of the following statements are true regarding previously cooked and cooled TCS foods (PHFs)?
   a. They must be rapidly reheated to an internal temperature of 165°F (74°C) for 15 seconds within two hours
   b. Food that does not reach 165°F (74°C) within two hours must be discarded
   c. Food that does not reach 165°F (74°C) within four hours must be discarded
   d. Food that does not get cooled to 70°F (21°C) within two hours must either be discarded or reheated to 165°F and cooled correctly

9. What are two food service controls that can be used to minimize the risk of contamination?
   a. Serve food quickly after cooking, cooling or reheating
   b. Separate raw food and ready-to-eat food
   c. Allow hot foods to cool to below 135°F (57°C) before serving
   d. Package leftovers in plastic containers

10. What are two steps employees can take to protect against food service contamination?
    a. Practice good personal hygiene
    b. Keep only a few serving utensils at self-serve buffets
    c. Prepare food in large batches
    d. Minimize handling of food

FACILITIES AND EQUIPMENT

FACILITY DESIGN

LESSON 1

After completing this lesson, you should be able to:

- Describe how the proper design, construction, and layout of your premises can reduce the risk of contamination
- Itemize the requirements for hand-washing stations
- Identify the guidelines for storing specific food items

Good Design

Good design and regular maintenance of food premises are essential to avoid hazards such as the contamination of food and multiplication of bacteria. The food premises must:

- Be large enough to accommodate all essential equipment
- Have separate areas for the storage and preparation of raw foods and time/temperature control for safety (potentially hazardous) foods
- Ensure a continuous workflow from delivery to service
- Ensure the separation of food and waste

Most importantly, the food premises themselves must be kept clean and in good repair, or all of these design benefits will be lost.

The most important design principle for food premises is that the workflow should be linear or continuous, progressing from raw material to finished product and ensuring the distances traveled by food or food handlers are kept to a minimum. This is vital to prevent contamination and cross-contamination, as it ensures minimal or no contact between raw foods and ready-to-eat foods.

Poor Design, Construction, Layout

An example of a poorly designed food premises is one that is located in an area prone to flooding, which will likely result in sewage contamination, or next to an operational factory, resulting in chemical contamination. An example of a poorly constructed food premises is one put together with materials that encourage condensation or having open joints or gaps where dirt and bacteria collect, resulting in contamination. An example of a food premises with a poor layout is one where raw food and ready-to-eat food are prepared together, which results in cross-contamination.

Suitable and Sufficient Facilities

Suitable and sufficient facilities are required for cleaning and sanitizing, washing and staff. This applies to both the food premises and the food equipment. You should always keep cleaning chemicals and materials separate from food.

Washing

Hand-washing stations are required in restrooms, dishwashing areas and food preparation and service areas. All stations must be fully functional and must also have hot and cold water, soap, drying method and trash receptacles. There must be a sign posted directing employees to wash their hands before returning to work.

Staff facilities may include lockers, restrooms, a break room and first aid equipment away from food, food storage and food production.

Restrooms

The design of the restrooms is especially important in food premises. Of particular importance is the location of the sinks for hand washing. The ideal place is between the toilet and the door. Restroom doors must not open directly into food preparation areas. It is vital that both staff and public restrooms are kept clean to avoid contamination and the spread of bacteria in the premises.

Temperature control

There should be sufficient equipment to enable the temperature control of food, such as refrigerators, freezers, cooking and cooling facilities. Refrigerators and freezers must be placed away from sources of heat, such as cookers and windows.

Pest Proofing

Food premises must be pest proof. This means preventing insects and animals from gaining access to the premises.

Approval and Compliance

If you are designing your facilities from the ground up, or are remodeling existing premises, you'll want to check with local regulatory agencies about approval of your design plans. Even if approval is not required, a review will ensure that you are in compliance with any existing regulations. If the review uncovers any flaws in your plans, resolving the issues before construction begins will save you time and money in the long run. Additionally, the Americans with Disabilities Act (ADA) outlines requirements for premises that apply to both employees and guests with disabilities. See the ADA Accessibility Guide (ADAAG) for these requirements.

After construction, you must obtain an operating permit. Before the permit is granted, your premises may have to undergo inspection to ensure that any required design specifications have been met. Once your premises have passed inspection, you'll be ready to open!

Floors

Flooring materials must be selected based on health and safety requirements. All flooring in the premises should be sturdy and easy to clean. Other important things to consider when selecting flooring are how well the flooring will wear over time and whether or not it is anti-slip. Anti-slip flooring, however, should be used only in high-traffic areas.

Nonporous, resilient flooring

Vinyl floor

Ceramic tile

Rubber mat

Areas of the food premises that need to use some kind of nonporous, resilient flooring include walk-in refrigerators, dishwashing areas, restrooms, garbage storage areas and other areas where the floor may be subject to moisture, flushing or spray cleaning.

Nonporous, resilient flooring is typically made of rubber or vinyl tile. It withstands shock well and is fairly inexpensive. Other characteristics that make it a good candidate for these areas are its durability, the fact that it's easy to clean, it's grease resistant and it's easily repaired or replaced. The only major caution about using nonporous resilient flooring is that it can be damaged easily by sharp objects.

Hard-surface flooring

Another type of flooring is hard-surface flooring. Hard-surface flooring is commonly made of ceramic tile, brick, marble or hardwood. Like nonporous, resilient flooring, it is also nonabsorbent and durable; however, this type of flooring is also more expensive and is not as easy to clean. The materials used in hard-surface flooring may cause breakage of objects dropped on it. It can also be slippery and does not absorb sound well. Hard-surface flooring is a good choice to use in restrooms and other high-traffic areas such as entrances and lobbies.

Carpet

According to the FDA *Food Code*, carpeting may not be used in walk-in refrigerators, dishwashing areas, restrooms, garbage storage areas or any other areas where the floor may be subject to moisture, flushing or spray cleaning. Typically, carpeting is the preferred flooring for use in dining room areas because it absorbs sound. Carpeted areas should be vacuumed daily or more often if necessary.

Nonslip flooring

Nonslip flooring should be used in high-traffic areas—especially the kitchen. It is acceptable to use rubber mats in areas where there is likely to be standing water, such as the dishwashing area. These mats should be picked up off the floor and cleaned, the floor should be mopped or scrubbed as usual and then the mats can be replaced.

Walls and Ceilings

The walls of the food premises should be sealed, sturdy and easy to clean. Ceilings should be covered, and joists and rafters should not be exposed to moisture. Any utility service lines and pipes should not be exposed unnecessarily. In addition to the walls themselves, any fixtures attached to the walls in food areas—such as light fixtures, vent covers and wall fans—must also be easily cleanable.

WATER

LESSON 2

After completing this lesson, you should be able to:

- List at least two sources of drinking (potable) water supplies
- Describe the best way to prevent backflow
- List three uses of non-drinking (non-potable) water permitted in a food service facility

> ## TERMS TO KNOW

AIR GAP The space between a water outlet and the highest level water can reach in a sink drain or tub. It is one of the cheapest and most reliable methods of backflow prevention.

BACKFLOW The reverse flow of water from a contaminated source to the drinking (potable) water supply. It can occur when there is a drop in water pressure and water from the contaminated supply is sucked into the drinking (potable) water system.

CROSS-CONNECTION The mixing of drinking (potable) and contaminated water in plumbing lines.

POTABLE WATER Water that is safe to drink; an approved water supply.

All food premises must have a satisfactory, constant supply of drinking (potable)water. Only **potable water** can be used in food preparation and for cleaning food and food contact areas. This requirement includes water used to clean equipment, in prep sinks and dishwashers and steam tables. Sources of potable water include a public water system; a private water system that is properly maintained, sampled and tested annually; bottled drinking water; properly maintained water pumps, hoses and pipes; water transport vehicles and water storage containers. Non-drinking (non-potable) water may only be used in a food establishment for air conditioning, non-food equipment cooling and fire protection.

Non-Potable Water

The only exceptions for using potable water in a food premises is water used for air conditioning, fire protection and non-food equipment cooling. The non-potable water must not come into contact with any food or food contact surfaces. Non-potable pipes must be labeled as such.

Private Water Supply Systems

If the food premises use a private water supply system, it should be inspected at least annually. Check with your local regulatory agency for specific information on testing and inspections. A copy of the most recent sample report should be kept and filed on the premises.

If the construction of the food premises is new, the water system should be flushed and sanitized before becoming operational to remove any possible contaminants. There are other times when flushing the system will be necessary. The system should also be flushed after a repair or modification to the system has been made and after an emergency situation has occurred, such as a flood.

Emergency Guidelines

If an emergency situation has occurred, such as a hurricane or flood, that has affected the potable water supply on the premises but there is a need to maintain service, there are guidelines to follow to ensure safety. Listen to announcements from local authorities that will advise whether the water is safe to use. If the water is deemed unsafe, boil it to kill harmful bacteria and parasites that may be present. Most organisms are killed in water that is held at a rolling boil for 60 seconds. But be aware that boiling will not remove chemical contaminants. If you are unsure whether or not the water has been chemically contaminated, use bottled water from a safe source. It is also possible to treat water with bleach, chlorine tablets, or iodine tablets. If you attempt any of these water treatment methods, check the label for directions on how to properly use the products. Be aware that many parasitic organisms will not be killed using these methods. Boiling is the best method to use.

The Plumbing System

The primary purpose of the plumbing system for the food premises is to prevent drinking (potable) water from mixing with non-drinking (non-potable) water. When the two types of water come into contact with each other, a **cross-connection** has occured. Cross-connections are known to have caused foodborne illness outbreaks.

An example of a cross-connection is when a hose is connected to a faucet dispensing drinking (potable) water and placed in a bucket containing contaminated water, such as water that was used to mop the floor. If a reverse flow occurs, the contaminated water in the bucket will mix with the drinking (potable) water coming out of the faucet. This occurrence is called **backflow.** Similarly, back siphonage is when there is a cross-connection and a reverse flow is caused by a sudden drop in pressure in the drinking (potable) water system that sucks the contaminated water back up into the drinking (potable) water lines.

There are several different types of approved backflow prevention devices available:

- Barometric loops
- Vacuum breakers (atmospheric and pressure types)
- Dual-check valves with intermediate atmospheric vents
- Dual-check valve assemblies
- Reduced pressure principle devices

If any of these devices are installed, they must be placed where they can be properly serviced and maintained according to the manufacturer's specifications.

Air Gaps

Although backflow prevention devices are available, the best way to prevent backflow from a cross-connection is the non-mechanical method: the **air gap**. An air gap is the air space that separates the opening to a drinking (potable) water supply line from a potentially contaminated source, such as a floor drain. An air gap between the water supply inlet and the flood level rim of the plumbing fixture, equipment, or non-food equipment must be at least twice the diameter of the water supply inlet and may not be less than 25 mm (1 inch).

Leaks

Another part of an establishment's plumbing system is overhead pipes carrying waste water. These pipes should be clearly marked non-potable. If these pipes leak, contamination of food and food preparation surfaces will occur. Even if the overhead pipes contain drinking (potable) water, a leak is still a problem. Condensation that forms on the leaky pipe can drip down onto food and food preparation surfaces. You should have all of the lines checked and serviced immediately when a leak has occurred.

Sewage and Waste Water

Food premises should have an efficient drainage system to remove sewage and waste water quickly without flooding. Sewage and waste water are highly contaminated and contact with food or food preparation surfaces must not occur. The drainage system should allow sufficient access for cleaning in the event of blockages and be constructed in such a way as to prevent pests from trying to live in the system. If grease traps are used on the premises, they must also allow for easy access and be maintained and serviced by a licensed plumber. Additionally, the drainage system must empty directly into the public sewage treatment facility or into another approved disposal system that operates according to legal specifications.

GUIDELINES AND MAINTENANCE

LESSON 3

After completing this lesson, you should be able to:

- List materials that are suitable for the construction of food service equipment
- Outline standards for placing and cleaning movable and stationary equipment
- Review the types of temperature-measuring devices available for kitchen use

Kitchen Equipment

Clean, well maintained kitchen equipment is a minimum standard, not an option, in the prevention of cross-contamination. Equipment that is designed for kitchen use and made of materials that will not collect dirt and bacteria is the easiest to keep clean.

Ventilation

In general, the ventilation should be sufficient to reduce room temperatures and humidity and maintain the overall indoor air quality of the food premises. A proper ventilation system helps remove vapors, odors and fumes that can cause contamination. If the food premises have proper ventilation, you will find little to no grease buildup on walls and ceilings.

Ventilation systems must be designed and installed according to legal specifications. At installation, ventilating systems, as well as heating and air conditioning systems, must allow for the intake of make-up air and have exhaust vents that do not cause contamination. Make-up air is extra ventilation that creates a healthier interior environment and ensures that plenty of fresh air is entering the premises. Make-up air intakes must be screened and filtered to prevent dust, dirt, insects and other contaminating material from entering the premises.

The ventilation system must be configured so that hoods, ductwork, and fans don't drip onto food or equipment. All hoods should be tested prior to use, and must be tight fitting, yet easily removable.

In a ventilation system, the purpose of an exhaust hood is to provide a way to collect all of the grease produced from the cooking process. It also serves as a means of removing heat, smoke and odors from the cooking area. For the hood to do its job, there has to be enough air movement to take the grease particles and cooking odors and pull them up to the grease extractors. This keeps the grease particles from clinging to nearby surfaces.

The ventilation equipment, including hoods and ductwork, should be cleaned regularly. Check with your local regulatory agencies to ensure that the ventilation system on the food premises meets all requirements and standards.

Good Lighting

Food premises must have adequate lighting so that food handlers can carry out their tasks safely and correctly and can identify hazards. Good lighting also helps ensure effective cleaning and sanitizing. Fluorescent tubes, fitted with protective sleeves or diffusers, as well as shatterproof bulbs and guards, are required where food could be exposed to risk from broken glass. Bright light also discourages pests from entering the food premises.

Minimum lighting levels

The FDA *Food Code* specifies intensity levels for the lighting in specific areas of the food premises.

- In areas such as dry storage areas and inside walk-in refrigerators and freezers, the minimum lighting intensity is 108 lux (10 foot candles).
- At buffets and bars, inside reach-in and under-the-counter refrigerators, above hand washing and dishwashing stations, in equipment storage areas, and in restrooms, the minimum lighting intensity is 215 lux (20 foot candles).
- Above food preparation surfaces where employees are working with knives, slicers, grinders, and other utensils, the minimum lighting intensity is 540 lux (50 foot candles).

Garbage

Careful consideration should be given to the storage and disposal of refuse, waste food and unfit food. Suitable receptacles should be provided both inside and outside the premises. Plastic garbage bags inside plastic or rubber bins are the typical method of handling the garbage inside the premises. Foot operated bins are recommended for internal use as lids are common vehicles of contamination. The bins should be emptied frequently into the external waste container and never allowed to overflow. When taking the inside garbage to the outside receptacles, do not carry the bags over food or food preparation surfaces to prevent potential contamination. Outside receptacles should be stored on concrete or asphalt, not dirt. Garbage receptacles for the outside are typically large metal garbage cans or dumpsters. External receptacles must be waterproof and leak proof, and must have tight-fitting lids. They should be made of metal or of a heavy-duty plastic and should be easily cleanable. The outside refuse area must also be kept clean. The area around the external receptacles should be hosed down regularly, including the drains, to avoid odors and deter pests. In addition, recyclables need to be stored in a separate area away from food.

Food Equipment

Food equipment that is made from inappropriate materials or that is cracked, chipped, broken, worn or badly designed is a haven for dirt and bacteria. It may also become a source of physical contamination or chemical contamination. Equipment that comes into contact with food should be smooth, waterproof, nontoxic, nonflaking, nontainting, resistant to corrosion, durable, easy to clean and suitable for its intended use.

Food equipment that cannot be dismantled, moved or easily cleaned is also hazardous. Examples of unsuitable materials include: soft wood, which is absorbent and therefore encourages cross-contamination and may splinter, leading to physical contamination; and materials such as copper, zinc, and aluminum, which may result in a chemical hazard, especially when used with acidic food.

Food equipment must be placed so that there is sufficient space to allow access to the equipment as well as access to all areas and other equipment around it. Where required, the equipment's placement must also allow for rapid dismantling and reassembly. The distance a piece of equipment is placed from a wall is determined by its size and ease of access for cleaning and maintenance. You should follow the manufacturer's specifications for equipment placement. Stationary equipment should be mounted on legs that are at least six inches off the floor. Likewise, stationary tabletop equipment should be on legs a minimum of four inches from the tabletop. Cracks or seams, wider than 1/32 of an inch where equipment is attached to a floor, wall or tabletop, must be sealed with a nontoxic, food-grade material. Sometimes equipment is designed so that it can be cantilever mounted, meaning that it can be attached to a wall or other surface with a bracket. This makes cleaning behind and underneath the mounted equipment much easier.

Whenever possible, equipment should be movable so that it is easier to clean. This is particularly important if the equipment is placed close to walls or other equipment. Regardless of being fixed or mobile, food equipment must be installed in such a way that allows the cleaning of its surrounding area.

Non-Food Contact Equipment

This equipment was not designed for food contact but is still used in food preparation areas. Non-food contact equipment has guidelines similar to those of equipment that does come in contact with food. These pieces of equipment may be splashed or spilled on or be soiled by food. Non-food contact equipment must be smooth, waterproof, corrosion resistant, easy to clean and maintain and simply designed without lots of ledges and hard-to-reach areas.

Purchasing Equipment

NSF International and Underwriters Laboratories (UL)

When choosing equipment for the establishment, use the list of approved equipment compiled by organizations such as NSF International and Underwriters Laboratories (UL). These organizations develop standards for sanitary equipment design and standards for environmental and public health. Approved equipment will be marked with the NSF or UL logos. Look for this mark on food service equipment before purchasing it! When choosing equipment for the establishment, be sure to use only commercial-grade food service equipment. The key reason for this is that common household equipment and appliances are not built for the heavy use seen by the equipment in a food establishment.

When considering the equipment needs of the food premises, there are several things you must keep in mind in addition to making sure the equipment is commercial grade:

- For large equipment such as sinks, dishwashing machines, refrigerators, and walk-in coolers or freezers, there are many sizes and styles from which to choose. Consider how much space you have to work with, how frequently the equipment will be used, how easy it will be to clean the equipment and the surrounding area, and how much routine maintenance will be needed to ensure proper operation. It is also a good idea to check with local regulatory agencies for specific requirements for this type of equipment.
- Consider the location of plumbing and supply lines when deciding on the type of equipment to purchase and its placement on the premises.
- After purchase and installation of the equipment, always refer to the manufacturer's specifications for proper use, maintenance, and cleaning procedures.

Utensils

As with other food contact equipment, utensils should be safe, corrosion resistant, waterproof, smooth, easy to clean and heavy enough to withstand frequent washing. Utensils should also be resistant to damage such as chipping, scratching and deformation. Utensils made of wood must be made of a close-grained hardwood, such as maple. Hardwood is an appropriate material for utensils such as cutting boards, rolling pins, salad bowls and chopsticks.

Rubber and rubber-like materials are acceptable for use as utensils provided they are durable enough to resist breaking, chipping and scratching. They must also be able to withstand being cleaned and sanitized in a commercial dishwasher.

While cast iron is not an acceptable material for utensils, it may be used as a food contact surface for cooking. Cast iron may also be used for serving food if the utensil is a part of the uninterrupted process that continues from cooking through service.

Once tableware and utensils have been cleaned and sanitized, they must be stored properly to keep them from becoming contaminated. Soiled tableware and utensils should be kept separate from clean items. The clean items should be kept in a clean, dry place that is at least 6 inches (15.24 cm) above the floor. They should not be exposed to food or dust. When presented to customers, the tableware should be prewrapped or stored so that only the handles are touched.

Thermometers

Choosing the right thermometer to monitor time and temperature is probably the most effective method you have to ensure that your food is safe. There are many different types of thermometers, and each type is designed with a specific use in mind. The thermometer most commonly used in food premises is a bi-metal stem thermometer. IR thermometers and thermocouples are also common. DO NOT use thermometers with glass sensors or stems unless the sensor or stem is encased in a shatterproof coating. The acceptable temperature range for a thermometer used in a food preparation area is 0°F to 220°F (–18°C to 104°C), and there is an allowed variance of +/– 2°F (1°C) for single scale thermometers.

Bi-metal stem thermometers have multiple uses. They are used to measure the temperature of received goods, as well as the internal temperature of food. A bi-metal stem thermometer measures temperatures using a metal probe with a sensing area near the tip. It can measure temperatures between 0°F and 220°F (–18°C and 104°C). Before using a bi-metal stem thermometer, make sure it meets the following criteria:

- The temperature markings are clearly numbered and easy to read.
- The accuracy is within 2°F.
- It has an adjustable calibration nut for accuracy.
- It is designed to undergo cleaning and sanitizing.

Thermometer calibration guidelines can be found on pages 126-127.

Thermocouples and thermistors measure temperature using a metal probe with digitally displayed results. They can be large or very small. There are other temperature measuring devices including immersion probes, surface probes, penetration probes and air probes. Infrared or laser thermometers are easy to use and deliver fast results. They use infrared technology to measure the surface temperature of both equipment and food. Because they measure food temperature without contact, the risk of cross-contamination is virtually eliminated. However, since they do not measure the internal temperature of foods, they must not be used for cooking, cooling or reheating of foods.

Maintenance Guidelines

It is not enough that food equipment is well designed and constructed. To avoid contamination, food equipment must be well maintained and used properly all the time.

Here are some guidelines for ensuring the good condition and proper use of equipment:

- Use all equipment according to the manufacturer's instructions. For example, always ensure that refrigerators and freezers are not overloaded, that they are at the right temperature, and that the doors are kept shut.
- Immediately report any damage that could cause the equipment—or the food in contact with it—to become unsafe or contaminated. For example, report any noticeable cracks or deterioration on door seals of refrigerators or freezers.
- Ensure that all equipment is effectively cleaned on a regular basis using the proper cleaning methods. This will avoid cross-contamination and the build-up of soil and dirt, which would otherwise allow bacterial multiplication and be an attraction to food pests.
- Never use the same equipment for handling raw and ready-to-eat food without thoroughly cleaning and sanitizing in between those food types. This will avoid cross-contamination.
- Use a color coding system to reduce the risk of cross-contamination by ensuring the same equipment is not used for different food types. Color coding, however, does not eliminate the need to clean and sanitize equipment. Color coding can be used for many types of equipment such as cutting boards, knife handles, work surfaces, cloths, protective clothing and packaging material.

Color coding systems may differ from one premise to another, so it is important that your staff knows and understands the color code in your workplace.

ASSESSMENT

1. Which of these is the most important concern with respect to the design of food premises?

   a. Inside location for all essential equipment
   b. Outside storage of refuse
   c. A continuous workflow from delivery to service
   d. A combined area for the storage and preparation of raw foods and ready-to-eat foods

2. Each of these is an example of suitable facilities for hand washing except:

   a. Separate sinks for food handlers dealing with raw food
   b. Separate facilities for hand washing and for food washing
   c. Several choices of soap at each sink
   d. Hand-washing sinks close to the entrance of the food preparation area

3. Where are hand-washing stations NOT required?

   a. Restrooms
   b. Break rooms
   c. Dishwashing areas
   d. Food preparation and service areas

4. According to the FDA *Food Code*, carpeting may be used in which of the following?

   a. Dining room
   b. Dishwashing areas
   c. Walk-in refrigerators
   d. Restrooms

5. Which of these is an acceptable way to use non-drinking (non-potable) water?

   a. Cleaning equipment and utensils
   b. Drinking
   c. Cleaning food contact surfaces
   d. Air conditioning

6. What is the best way to prevent backflow from a cross-connection?

   a. Barometric loops
   b. Air gaps
   c. Reduced pressure principle devices
   d. Double-check valve assemblies

7. All kitchen equipment must be:

   a. Movable
   b. Waterproof
   c. Sanitized daily
   d. Durable

8. Equipment not designed for food contact but used in food preparation areas should be:

   a. Smooth
   b. Waterproof
   c. Easy to clean and maintain
   d. All of the above

9. Which material may be used for serving food only if the utensil is a part of the uninterrupted process of cooking through service?

   a. Cast iron
   b. Hardwood
   c. Rubber and rubber-like materials
   d. Copper and galvanized metals

10. Which best describes the result of choosing the right thermometer to use?

    a. Keeps employees comfortable
    b. Ensures compliance with local regulations
    c. Promotes return customers
    d. Ensures that food is safe

CLEANING
AND SANITIZING

CLEANING VS. SANITIZING

LESSON 1

After completing this lesson, you should be able to:

- Explain the difference between cleaning and sanitizing
- Describe the guidelines for sanitization

> ## TERMS TO KNOW
>
> **CLEANING** The process of removing soil, food residues, dirt, grease and other objectionable matter; the chemical used to do this is called a detergent.
>
> **SANITIZE** To use chemicals or heat to reduce the number of microorganisms to a safe level. It is important to note that disinfectants and sanitizers are not the same thing. Disinfectants are not permitted in food preparation areas.

Cleaning and sanitizing are two separate steps that are an important part of reducing microorganisms in a food establishment.

The purpose of **cleaning** is to remove food residue, dirt and grease and other types of soil from surfaces.

You can do this by:

- Physical cleaning—for example, scrubbing
- Thermal cleaning—for example, the use of hot water
- Chemical cleaning—for example, the use of a detergent
- Combination of all of these

Generally, better chemicals and hotter water will mean that less physical cleaning will be required.

The purpose of **sanitization** is to reduce the number of microorganisms to a safe level. Microorganisms are destroyed by the use of very hot water, steam or a chemical sanitizer.

In order for this process to be effective, the equipment or surface must be thoroughly cleaned and rinsed before being sanitized, and the sanitizers need to be in contact with that equipment or surface for sufficient time, called the contact time.

All equipment and surfaces need cleaning at a certain frequency, but only some need sanitizing.

Sanitization is normally restricted to:

- Food-contact surfaces, such as cutting boards and knives
- Hand-contact surfaces, such as taps or refrigerator door handles
- Cleaning materials and equipment

If these surfaces are not sanitized, then the staff will be unnecessarily introducing food safety hazards to the operation. However, if equipment or surfaces are needlessly sanitized—for example, non-food contact surfaces, such as floors and ceilings—then food handlers will be wasting time and money.

It is also important that the staff know and understand which are the right cleaning chemicals and the right cleaning equipment for each task and how to use them. This will prevent physical and chemical contamination of equipment, surfaces and food.

CLEANING AGENTS

LESSON 2

After completing this lesson, you should be able to:

- Describe the factors that determine what kinds of cleaning agents should be used
- Identify the different types of cleaning agents

> **TERMS TO KNOW**

CLEANING AGENT A chemical compound, such as soap, that is used to remove dirt, food, stains or other deposits from surfaces.

DETERGENT A chemical or mixture of chemicals made of soap or synthetic substitutes. It facilitates the removal of grease and food particles from dishes and utensils and promotes cleanliness so that all surfaces are readily accessible to the action of sanitizers.

Cleaning Agents

Many factors can have an effect on how cleaning is done and what agent is most effective. Some of these include:

- The type of material being cleaned. This may determine what **cleaning agent** is appropriate. Some cleaning agents can damage certain surfaces.
- The hardness or softness of the water. Each type of water can cause a cleaning agent to react differently. Hard water may also cause a buildup of minerals that may need to be removed using a specific cleaning agent.
- The amount and type of soil that needs to be cleaned. Large amounts of caked-on soil will take more time and work to remove.
- The temperature of water being used. Some cleaning agents work better with the water at a specific temperature.

With any cleaning agent, you should read and follow the EPA-registered label use instructions. Make sure to check whether the cleaning agent is appropriate for use on a food-contact surface.

Cleaning agents can be less effective if used improperly, and mixing cleaning agents can cause toxic fumes. Many of these cleaning agents are considered hazardous:

- **Detergents**, which remove dirt and grease.
- Solvent cleaners, which help dissolve dirt and remove it. The solvents most commonly used in the restaurant industry are known as degreasers.
- Acid cleaners, which are used to remove tarnish, alkaline discoloration and corrosion from metals. They can also be used to remove hard-water deposits from dishwashers and serving tables.
- Abrasive cleaners, which are used to physically scrub dirt and soil from surfaces. Abrasive cleaners can damage surfaces and make them hard to clean in the future, so care should be taken when you choose this type of cleaner.

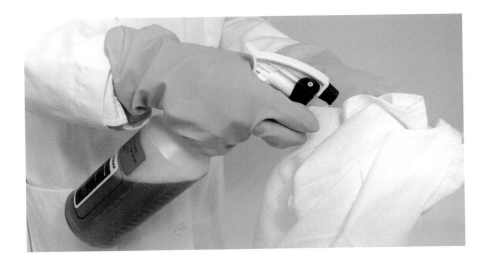

SANITIZATION

LESSON 3

After completing this lesson, you should be able to:

• Identify when sanitization is necessary

• Discuss the different methods of sanitization

It is essential to clean and sanitize all food preparation surfaces in a food establishment.

You should make sure all traces of the cleaning agent you used are removed before you sanitize your surface. Cleaning agents left on a surface could contaminate the food you are preparing or cause the sanitizer not to work properly.

When in contact with TCS food (PHF), most surfaces need to be cleaned and sanitized throughout the day or at least every four hours. Equipment, food-contact surfaces and utensils should also be sanitized:

• When you change to a different type of raw animal food

• When you change from working with raw foods to working with ready-to-eat foods

• When you change from raw fruits and vegetables to time/temperature control for safety (potentially hazardous) foods

• Before you use or store a food temperature measuring device

• At any time during the operation when contamination may have occurred

You must also have a procedure in place for employees to follow addressing specific actions to take responding to and minimizing the spread of contamination and exposure of employees, consumers, food and surfaces after vomiting or diarrheal events.

Heat Method

Hot water is one of the most commonly used sanitizers. It can be used when you are cleaning manually or by machine, and each scenario presents its own challenges.

When using hot water to manually sanitize objects, you must immerse the object for at least 30 seconds in water that is at least 171°F (77°C). You may need to add a heating element to your sink to maintain the correct temperature.

High-temperature dishwashing machines also use hot water (180°F [82°C]) to sanitize objects. The temperature of the water should be closely monitored to make sure that it is working properly. If the water is too hot, it could vaporize before it sanitizes the objects.

Chemical Method

Sometimes, using the heat method is not appropriate for sanitizing an object. Chemicals can also be used for sanitizing when cleaning manually or by dishwashing machine. The sanitizing chemicals must be approved for use on food-contact surfaces.

The chemical sanitizers most commonly used in the food safety community are:

• Chlorine

• Iodine

• **Quaternary ammonium compounds**, also known as quats.

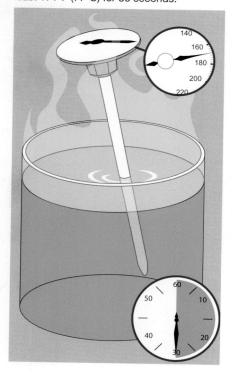

To use the heat method of sanitization an object must be immersed in water that is at least 171°F (77°C) for 30 seconds.

For chemical sanitizers to work effectively, several guidelines must be followed:

- Cleaning chemicals must always be used according to the EPA-registered label use instructions, company procedures and your instructions.
- They must always be used at the recommended concentrations and must never be mixed or diluted. A sanitizer test kit is used to check the concentration of the sanitizer, and each type of sanitizer requires a different type of test kit.
- Sanitizers must also be used at the proper temperature and be in contact with the surface for the required amount of time.

A chlorine solution must have a minimum temperature based on the concentration and pH of the solution as listed in the following chart:

Concentration Range	Minimum Temperature	
mg/L	pH 10 or less °F (°C)	pH 8 or less °F (°C)
25-49	120 (49)	120 (49)
50-99	100 (38)	75 (24)
100	55 (13)	55 (13)

Dishwashers

Dishwashers can help you save time by cleaning and sanitizing many items at the same time. You should always follow EPA-registered label use instructions when operating your machine and the machine should be kept in good repair.

- The correct water temperature for washing and sanitization should be monitored and maintained.
- Food and soil should be rinsed from utensils and objects before placing them into the dishwashing machine.
- All of the surfaces of the items should be arranged so that the water contacts the entire surface and items should be checked to make sure all food and soil have been removed when the machine has finished cleaning.
- The machine should be cleaned at least once every 24 hours, but it should be monitored throughout the day and cleaned as needed.

Commercial dishwashing machines are available in two types:

- High-temperature machines

High-temperature machines use hot water and detergent to clean and hot water to sanitize. The temperature of the wash cycle in a high-temperature machine can vary according to the type of machine you are using. You should check the EPA-registered label use instructions and the FDA *Food Code* to determine the proper wash temperature for your machine. The final rinse temperature should be 180°F (82°C) to ensure sanitization. If the final rinse temperature is too hot, over 195°F (91°C), the water will vaporize before the items are sanitized.

- Chemical sanitizing, or low-temperature, machines

Chemical machines operate at a lower temperature than the high-temperature machines and use a chemical sanitizer. The temperature that the machine operates at will be determined by the type and concentration of chemical sanitizer you use, but It should not be lower than 120°F (49°C) for the wash cycle.

An iodine solution must have:

- A minimum temperature of 68°F (20°C)
- A pH of 5.0 or less or a pH no higher than the level for which the manufacturer specifies the solution is effective
- A concentration between 12.5 mg/L and 25 mg/L

A quaternary ammonium compound solution must:

- Have a minimum temperature of 75°F (24°C)
- Have a concentration as specified under § 7-204.11 and as indicated by the EPA-registered label use instructions included in the labeling
- Be used only in water with 500 mg/L hardness or less or in water having a hardness no greater than specified by the EPA-registered label use instructions

THE SINK

LESSON 4

After completing this lesson, you should be able to:

- Explain the proper dishwashing station setup
- Describe the proper procedure for washing items manually

Items that cannot be washed and sanitized in a dishwashing machine must be washed manually.

A three-compartment sink is the most common type of sink used. Two-compartment and four-compartment sinks are sometimes used as well. Check the regulations for your area for specific information.

The compartments of the sink should be large enough that you can fully immerse any item that needs to be washed by hand.

Dishwashing Station Setup

Your manual dishwashing area should be set up with specific arrangements in mind:

- You need an area to store dirty and soiled items and a separate area to air dry the clean and sanitized items.
- A separate area should be used to remove food from items before they are washed.
- Your station should include a hand sink, so that hands can be washed between handling dirty and clean dishes. It should also include a thermometer to measure water temperature and a clock with a second hand to monitor sanitization times.
- If you are using a chemical sanitizer, you need to have test kits available to check the concentration of your sanitizing solution.

Proper Procedure

When washing items manually, you should:

1. Start with a clean and sanitized sink and dishwashing station. You should remove as much of the visible soil and food as you can before washing in the three-compartment sink.

2. Use the first compartment to wash the items. Use a detergent. Make sure that the water temperature used is at least 110°F (43°C). Replace the detergent and the water in this compartment as needed.

3. The second compartment is used to rinse the detergent off of the item and prepare for sanitization. You can spray clean the objects or immerse them in water to remove the detergent. The water used in this compartment should be kept clean and replaced when needed.

4. The third compartment holds a sanitizing solution. It may need a device to keep the temperature of the sanitizing solution at a specific temperature. The temperature of the solution and the contact time for the solution will be determined by the type and concentration of chemical sanitizer you use.

5. For hot-water sanitization, you must immerse the object for at least 30 seconds in water that is at least 171°F (77°C).

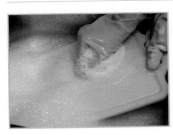

6. Once an item is sanitized, it should be allowed to air dry.

CLEANING IN PLACE

LESSON 5

After completing this lesson, you should be able to:

- Identify clean-in-place (CIP) equipment
- Demonstrate a typical CIP sequence
- Define the essential benefits of proper cleaning
- Identify ways to avoid contamination

> **TERMS TO KNOW**

CIP Stands for cleaning in place. This cleaning process is necessary when equipment cannot be dismantled or moved. It involves the circulating of nonfoaming detergents and disinfectants, or sanitizers, through assembled equipment and pipes, using heat and mostly turbulence to attain a satisfactory result.

Food premises should be designed and constructed in such a way as to promote and enable effective cleaning and sanitizing. Sometimes this is not a straightforward process.

CIP—or cleaning in place—is necessary when equipment cannot be dismantled or moved. This process is most common in dairy premises, breweries and potable-liquid manufacturing or when the equipment is large, floor mounted or sealed. CIP equipment should be self-draining or capable of being completely drained of cleaning and sanitizing solutions.

Equipment that cannot be disassembled should have access points that allow you to inspect the equipment to make sure that it is properly sanitized.

Prior to cleaning CIP equipment, unplug electrical equipment and remove detachable parts for manual wash.

A typical CIP sequence consists of five steps:

1. Pre-rinse, to remove soil in the pipes
2. Detergent circulation, to remove residual debris and dissolve grease or soiling
3. An intermediate rinse with water
4. Sanitization, to destroy the remaining organisms to a safe level
5. Air dry

Reassemble any parts removed for manual washing and resanitize parts handled during reassembly. You should follow all EPA-registered label use instructions and the FDA *Food Code* for all clean-in-place equipment. You should take extra care with refrigerators and dangerous equipment such as meat slicers.

Soiled Surfaces

The soiling of surfaces and equipment is unavoidable in food businesses. However, dirty premises will create a poor working environment and lead to customer complaints. It will also introduce food safety hazards. Examples of this include:

- Contamination of equipment and food by waste, unfit food, packaging materials, dust or soil
- Contamination of equipment and food by food pests, attracted to the food premises by wastefood, rubbish and odors
- Contamination of equipment and food by bacteria, including cross-contamination of potentially hazardous foods
- Multiplication of bacteria, as the conditions required for their growth are not being removed

On some occasions, unsatisfactory or negligent cleaning may be an even greater hazard than not cleaning at all, as it may result in the redistribution of physical and bacterial contaminants within the workplace. It may even result in the premature replacement of equipment or part of the premises, for example, due to corrosion.

Cleaning the establishment is essential to:

- Safe and pleasant food premises
- Removing contaminants
- Preventing food spoilage, food poisoning and foodborne diseases
- Preventing damage or reduction in the efficiency of food equipment
- Complying with the law

Non-food-contact surfaces should be kept clean and free from debris.

Cleaning tools are often vehicles of physical, chemical and bacterial contamination. To prevent contamination:

- All tools that clean food contact surfaces should be cleaned and sanitized after use.
- Where appropriate, tools should be left to air dry.
- Staff should never leave cleaning materials in dirty buckets or soaking in water overnight.
- Just like cleaning chemicals, cleaning tools should be kept separate from food items.
- To prevent cross-contamination, there should be tools used for time/temperature control for safety (potentially hazardous) food areas only and other tools for raw food areas.
- Alternatively, ready-to-eat food areas must be cleaned before raw food areas.
- All areas must be cleaned from top to bottom.
- Color coding of tools is recommended as an effective way to reduce the risk of cross-contamination. That way, for example, tools used to clean the restroom aren't used to clean the meat slicer.

Cleaning supplies, such as brooms, mops and wiping cloths, should be stored away from any food-contact surfaces. All should be air dried unless they are stored in a sanitizing solution. If a cleaning tool is used in a non-food-contact area, you should use a different, sanitized cleaning tool for the food-contact areas.

HAZARDOUS MATERIALS

LESSON 6

After completing this lesson, you should be able to:

- Recognize the importance of material safety data sheets (MSDSs)

- Illustrate how hazardous chemicals should be stored

Hazardous Materials

Some materials essential to a food establishment are hazardous. To avoid harm it is imperative that specific standards are met.

OSHA Requirements

It's a fact that there are hazardous materials in the workplace, but only the hazardous materials required for operation are allowed in an establishment serving food.

Detergents, sanitizers, pesticides and other chemicals can be harmful when not used properly. The Occupational Safety & Health Administration, or OSHA, sets standards for the use of hazardous materials in the workplace. Some states have created OSHA-approved plans and set their own standards. Check with your local agency to make sure you are in compliance.

Regardless of where the plan originates, you must follow the Hazard Communication Standard, also known as HCS or HAZCOM. The HCS provides employees the right to know what hazardous materials are in their workplace and how they should handle them.

MSDS Details

A vital part of the HCS is the availability of material safety data sheets, or MSDS. These are created by the manufacturers of the chemicals and contain information about the hazards of the specific chemicals and directions for safe use. OSHA requires that the MSDS for each chemical be available to all employees in their work area. All employees should be trained on how to read material safety data sheets.

Storing Hazardous Materials

OSHA requires that manufacturers label their containers with:

- Their name and address
- The chemical in the container
- The appropriate hazard warnings for that material

If you put the chemical into another container, you must label it with the same information.

Chemicals must be stored in such a way that they will not contaminate food, equipment or other objects. This means that they must be placed in a separate area or in a place that is not above the food and equipment, not only in food prep areas, but also in storage areas.

DESIGNING A CLEANING PROGRAM

LESSON 7

After completing this lesson, you should be able to:

- Explain the importance of maintaining a cleaning schedule
- Illustrate what should be specified for each task
  on a cleaning schedule

Designing a Cleaning Program

The best way to ensure that your food establishment is being cleaned properly is to create a cleaning schedule. To ensure that cleaning is effectively implemented throughout all areas of a food business, cleaning needs to be planned and organized, and all staff must know and understand their responsibilities.

The easiest way to do this is to create a written cleaning schedule (SSOP— Sanitation Standard Operating Procedure).

Steps

Cleaning schedules must clearly specify details regarding the surface or equipment to be cleaned, when it is to be cleaned, and the staff who should be involved in cleaning.

It's not enough that cleaning is done as stipulated in the schedules. In addition to following a cleaning schedule, staff should also adopt a clean as you go policy to prevent dirt from accumulating—for example, mopping up spills immediately. This will also contribute to a safer work environment as it will prevent slips, trips and falls.

Tasks

Sanitation standard operating procedures (SSOPs)

For each surface and piece of equipment you need to provide standard operating procedures that specify:

- What is to be cleaned
- How it is to be cleaned or sanitized—that is, the method to follow
- How much time is to be allowed for the task
- What standard is required
- What chemicals to use, including the amounts needed
- What contact time should be allowed
- What equipment should be used, including references to color coding
- Where and how to store chemicals and equipment after the task

Finally, the cleaning procedures must also specify what safety precautions are to be taken, including any appropriate protective clothing or measures to protect food such as covering all open food and other surfaces, and who is responsible for monitoring and recording that the equipment has been cleaned to the appropriate standard, including checking that any dismantled or moved equipment is now safe to use.

Monitoring

It is the manager's responsibility to produce cleaning schedules or delegate this task to an appropriately trained staff member. The success of your program depends on several factors. All of your employees need to be aware of and trained on the cleaning plan and how it works.

ASSESSMENT

1. What is the purpose of cleaning?

   a. To reduce the level of microorganisms
   b. To remove food residue, dirt and grease from surfaces
   c. To give staff something to do with their free time
   d. To test new cleaning agents available on the market

2. What is the term for the process of reducing microorganisms to a safe level?

   a. Chemical cleaning
   b. Thermal cleaning
   c. Physical cleaning
   d. Sanitizing

3. Which of the following statements is true about cleaning and sanitizing?

   a. All equipment and surfaces need to be cleaned
   b. All equipment and surfaces need to be sanitized
   c. All equipment and surfaces need to be cleaned and sanitized
   d. Equipment that is not cleaned properly must be sanitized

4. When surfaces are in contact with time/temperature control for safety (potentially hazardous) foods, what is the maximum amount of time allowed between sanitizations?

   a. 12 hours
   b. 8 hours
   c. 4 hours
   d. 1 hour

5. Each of the following statements is true regarding dishwashing machines except:

   a. The machine should be cleaned at least once a week
   b. The correct water temperature should be monitored and maintained
   c. Items placed in the dishwasher should be rinsed first
   d. The manufacturer's operating instructions should be followed

6. For hot-water sanitization, items should be immersed in water for 30 seconds at what temperature?

   a. 185°F
   b. 171°F
   c. 121°F
   d. 98.6°F

7. Which two of the following statements are true of CIP equipment?

   a. It should have sufficient access points to allow inspection
   b. It should be professionally cleaned
   c. It needs to be cleaned at least once a year
   d. It should be self-draining

8. What is the order of a typical CIP sequence?

   a. Wash with detergent, first rinse, inspect, second rinse, air dry
   b. Rinse, wash with detergent, sanitization, rinse, air dry
   c. Pre-rinse, wash with detergent, intermediate rinse, sanitization, air dry
   d. Pre-rinse, sanitization, intermediate rinse, wash with detergent, final rinse

9. What resource would you check for information about the hazards of a specific chemical and directions for safe use?

   a. The OSHA log book
   b. The FDA *Food Code*
   c. The MSDS
   d. The HAZCOM plan

10. Cleaning schedules must clearly specify details regarding all the following except:

    a. Who should be notified if a cleaning supply needs to be replenished
    b. What equipment should be cleaned
    c. Which staff members are responsible for what tasks
    d. When the cleaning should take place

CHAPTER EIGHT

PEST CONTROL

INTEGRATED PEST MANAGEMENT

LESSON 1

After completing this lesson, you should be able to:

- Describe the benefits of a pest control program
- Identify where pests are most likely to enter and remain on the premises
- List best practices for pest control

> **TERMS TO KNOW**
>
> **PEST** An animal, bird or insect capable of directly or indirectly contaminating food.

Pest Control

Food pests can have a dramatic effect on the reputation and profitability of food service facilities, not to mention the health of workers and customers. Some pests look so disgusting that once seen in a restaurant, a customer will never return again. Other pests are far more serious. They can enter food, spreading disease and illness, even death. The job of pest control is to prevent such pests from entering a facility and, if found, to safely eliminate them.

Common **pests** found in the food industry include:

- Rodents, such as rats and mice
- Insects; for example, flies, wasps, moths, cockroaches, booklice, silverfish and ants
- Birds; for example, pigeons and sparrows, especially in outside eating areas
- The occasional small animal, such as a stray cat, around the dumpster

Pest control is essential to avoid:

- Contamination and food waste
- Foodborne illnesses and foodborne disease outbreaks
- Damage to equipment and the premises in general—for example, fires caused by gnawed electric cables
- Loss of customers and profits caused by selling contaminated food
- Loss of staff who do not wish to work in infested premises
- Fines and closure of the establishment

Successful Implementation

Integrated pest management, or IPM, is an approach to pest control in which a wide range of practices are used to prevent or solve pest problems.

Successful implementation balances prevention with control. Prevention works to keep pests out—it is the key; control removes those that may get in. IPM recognizes that reliance solely on pesticides to eliminate pests after they enter an establishment can permit an infestation to get out of hand before it is even noticed. The strength of IPM is in preventing an occurrence in the first place.

Goals

Successful implementation also requires the creation and communication of goals. Goals help inform employees of their responsibilities under the facility's IPM and convey the actions needed to prevent infestations. All IPM plans have three basic goals:

1. Prevent pest entrance to the facility

2. Remove food and water sources, as well as living areas, shelter and harborage areas to eliminate and prevent breeding
3. Work with a licensed pest control operator (PCO) to control any infestations that occur

Many of the same pest prevention and control practices that apply to indoor eating also apply to outside areas. (On the other hand, bug zappers are typically found only in outside areas.) For this reason, outside areas also require pest prevention and control practices.

Pest Entrance

Pests enter buildings in two ways. The first is through gaps, holes or openings in the building. Examples of openings are a crack in the wall, a gap between walls and window frames, and through a door left open. Second, pests can be brought in by people, supplies or even equipment entering or leaving a building. Food delivery is of the greatest concern in this regard.

Walls, floors and ceilings

To prevent access through the building structure itself:

- Seal all wall openings. Notice light or feel air coming through a wall. Correct with a sealer, one that is permanent and, if necessary, can adjust to outside temperature changes.
- Patch floor cracks and holes. Look for hairline openings in the floor, especially around uneven areas. Correct with a patch that both seals and can withstand the weight of the people and equipment travel.
- Repair ceiling leaks. Follow a drip or water mark to its source—water can travel for some distance along rafters and trusses before becoming visible. If the leak is from an outside roof, have a professional inspect and maintain roofing materials. If coming from a floor above the facility, contact building maintenance.

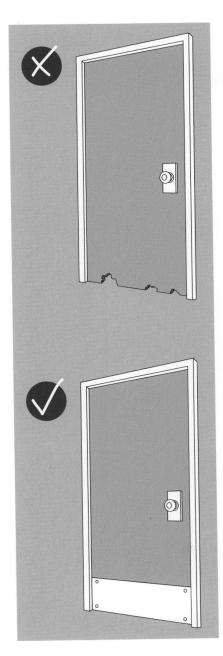

Doors and windows

To prevent access through doors and windows:

- Maintain door seal. Check for correct door fit. Maintain weather stripping for a tight seal on door top and sides. Install a door sweep to seal door bottom.
- Keep doors closed. Post signs to remind workers, as well as customers. For doors that are frequently left open, install automatic closing devices.
- Consider an air curtain, if appropriate. An air curtain creates a bug barrier by directing a flow of air away from the door, pushing flying bugs back and out of the facility.
- Inspect window seals. Install a sealant in gaps between door and window frames and walls.
- Install and maintain screens on doors and windows that open. The size of the screen is important. For doors and windows that are kept open for ventilation, the FDA *Food Code* specifies that the required screen mesh is 16 mesh to 25.4 mm (16 mesh to 1 inch) to protect against entry by insects and rodents.

Plumbing and HVAC

To protect against access through plumbing and heating, ventilation and air conditioning (HVAC) systems:

- Eliminate leaks and drips. Pests thrive in moist areas. Fix leaky faucets, dripping or sweating pipes and leaks in water or drainage piping.
- Inspect dishwashing machine and hot water heaters for leaks. Also, check fresh water and drain pipes, especially if rubber or synthetic material is installed.
- Guard floor drains. Cover floor drains with a removable grate or strainer. Immediately clear plugged drains.
- Seal gaps around pipe and duct work that passes through walls. Select a sealant appropriate to the purpose—synthetic, metal or concrete. Your PCO can provide guidance.
- Install screens over external ventilation vents and ducts. As with screens for windows and doors used for ventilation, install 16 mesh screens. Clean often.

Food delivery

To protect against access through food delivery:

- Select an approved, reliable supplier. Your suppliers should be sensitive to food service pest control concerns.
- Inquire into a supplier's pest control plan. What do the suppliers do to assure stock rotation? Are pesticides used in storages areas? How is spoilage handled? How are pests prevented from entering trucks or the building at the loading dock?
- Inspect deliveries. Inspect truck for cleanliness. Open crates and cartons. Look for signs of pests, such as odor, droppings, dead pests and so forth.

When inspecting and correcting building access issues, consult with your PCO or local health organization. Each has a wide knowledge of identification and correction techniques to support your pest prevention efforts.

What Attracts Pests

Pests are attracted to places where food and water are readily available. Additionally, they find dimly lit, moist, warm locations good places to live and breed. What they don't like is just the opposite. Pests will not thrive in dry, well-lit areas where food is scarce.

Facility cleanliness

The cleaning schedule for the facility should include:

- Areas or items to clean. Include floors and drains, walls, ceilings, HVAC, doors, windows, equipment, work surfaces, toilets and rest areas, dining areas and so forth.
- Precautions to take for each area or item.
- Spot inspections to monitor the effectiveness of the schedule. Are scheduled times working? Are more frequent inspections necessary? Fewer inspections? Are there other things that need inspection?

Refuse and waste disposal

The refuse and waste disposal cleaning schedule should include:

- Inside and outside waste containers. Pests and wildlife will have no trouble finding a food source at soiled or overflowing containers. Dispose only of what the container can hold. Check for a lid that fits tightly. Clean frequently, cleaning up any spillage on the ground.
- Food preparation areas. Clean up work area frequently. Discard tainted food immediately.

- Other containers. Bottles and cans provide food and excellent housing for pests. Keep a tight lid on recyclable or returnable containers. Rinse out cans and bottles, if practical. Remove cardboard boxes once they are emptied. Keep worn food preparation clothing and reusable rags in a sealed container. Clothing used during food preparation also provides a food source and shelter.

Other Priority Practices

In addition to a cleaning schedule for routine tasks and for refuse and waste disposal practices, other priority practices that eliminate food sources include:

- Good housekeeping in employee locker and break areas. Food on worn clothing is a magnet for pests. Put stained clothing in sealed, used laundry hampers. Keep snack containers and drink cups picked up.
- Immediate spill and waste cleanup, both in food preparation and food serving areas. In addition to pest prevention, debris on food serving floors is a signal to customers as to the general cleanliness of the entire facility.
- Frequent restroom checks during busy serving hours.
- Rinsing and storing floor cleaning materials after use. Empty mop buckets immediately after use. Rinse out mops with clean water. Dry and store cleaning tools after use.

Outside Dining Areas

Outside dining areas provide their own distinct challenges to pest control. Unlike indoor areas, there are no barriers, such as walls, doors and windows to keep pests out. Ants, bees, flies and even birds have free access. Squirrels are not the only wildlife that patrons can feed. The challenge in outdoor eating areas is to minimize pest incursions rather than achieve complete prevention. To minimize pests in outdoor eating areas:

- Bus and clean tables immediately. Waste food and drink on tables is a sign to come right in. Pick up and remove plates and glasses. Clean table tops and inspect for spilled food or drink on floors and seats. Keep condiment containers clean and sealed.
- Maintain landscaping. Trees, bushes and ground cover are all natural hiding places for pests. Keep grass mowed, bushes trimmed and weeds under control. If possible, create a gravel or concrete space between the outside of the eating area and landscaping or lawn.
- Keep the immediate area dry and picked up. Water is a natural for attracting pests, especially mosquitoes. Eliminate standing water. Pick up papers and other refuse.
- Give special attention to outside dumpsters and garbage containers. Relocate dumpsters as far away and as down wind as possible. Contact your refuse companies for pressure cleaning of dumpsters. Keep all garbage cans clean and sealed tightly—place dirty tableware inside, behind a closed door.
- Discourage customers from feeding wildlife. Feeding wildlife creates an open-ended invitation for birds to find ALL their food at the facility. Explain the problem to the customer.
- Contact a PCO for additional help in minimizing the effect of pests in an outside dining area. A PCO can help you remove nests and hives. Further, since bug zappers, traps and other pest control tools can be helpful in reducing the occurrence of pests, a PCO can provide valuable assistance to placing and maintaining this equipment.

IDENTIFYING PESTS

LESSON 2

After completing this lesson, you should be able to:

- Identify the kinds of pests that cause the greatest risks to food safety

Pests that carry and spread disease are of the highest concern for pest control. Cockroaches, rodents and flies are of special concern because of their ability to cause illness. For this reason it's important to understand why they are of such high concern, how to identify them and where to look for them.

Cockroaches

How to identify

There are a number of different species of cockroaches, all of which range in size between ½ inch to 1¾ inches long. They are golden brown to dark brown to black in color. Their bodies are flat and broad with long antennae. They dislike light, prefer to be out at night and will scatter if disturbed. They have a strong, oily odor. Their feces look like specks of black pepper. Egg casings are also a sign of their presence.

Where to find

Cockroaches prefer undisturbed warm, moist places. They are found around sinks, drains, moist boxes and bags and wet flooring. Homes inside equipment, molding strips and even furniture are common. For kitchens with wood burning stoves, they are found in the wood. Regardless of the place, finding cockroach feces and crushed body parts is also another sure sign of their presence.

Common illnesses caused

They have been identified as carriers of salmonella and the poliomyelitis virus. They are responsible for food poisoning, dysentery and diarrhea. Their residue is allergenic, causing asthma symptoms in those sensitive to their allergens.

Rodents

How to identify

Mice are small mammals with hairless tails and large ears. Their color ranges from light to dark brown. Rats and mice have many of the same features, but differ in size. Rats are larger, up to nine inches in length. Both are typically nocturnal, good climbers and jumpers, as well as good swimmers. Feces from mice are dark brown and rice-shaped. Under black light, rodent pathways will appear as fluorescent trails.

Where to find

Rats and mice are gnawing animals that can chew entry through most materials (they are relatives of the beaver). For this reason, look for small, round entry holes near food and water supplies. Mice can squeeze through a hole as small as one quarter of an inch. Also, look for signs of nests (shredded paper, cardboard, cloth, or fur) and tracks across dusty surfaces. The surest sign of mice and rats is their feces, the rice-shaped pellets.

Common illnesses caused

They are known to spread Lyme disease, which typically starts with a rash and may progress to bouts of arthritis with severe joint pain and swelling. They also cause

Salmonellosis, which presents with gastrointestinal symptoms of diarrhea, fever and abdominal cramps 12 to 72 hours after infection, and leptospirosis, which has flulike symptoms. They're also known to carry the hantavirus, which is rare, but starts with flu symptoms and can progress to kidney failure.

Other Pests

Ants and flies, both common enough, pose very different health concerns.

Concerns about ants

Ants clean themselves regularly, especially their feet and antennae. Because they seldom carry debris from one place to the next, they are not generally considered disease carriers. They like to eat sweets, like fruit and sugar, so controlling ants is as simple as maintaining good housekeeping.

Concerns about flies

Flies, on the other hand, are a notorious health concern. Unlike ants, flies' bodies are covered with debris. If a fly has been in animal waste just before entering a facility, waste will be brought into the facility. The unhealthiest habit of flies, however, is their practice of vomiting on food before eating. Since flies can only ingest liquids, vomit acts to dissolve the food. Once dissolved, the food can then be ingested. Keep flies out by keeping doors and windows closed or protected with screens, or by installing air curtains to prevent entry.

Service Animals

Service animals that are controlled by a disabled employee or person may enter the food establishment in areas that are not used for food preparation and that are usually open for customers, such as dining and sales areas, as long as a health or safety hazard will not result from the presence or activities of the service animal.

PEST CONTROL

LESSON 3

After completing this lesson, you should be able to:

- Describe the role of pest control operators
- Identify different kinds of pesticides
- Describe the proper use and storage of pesticides

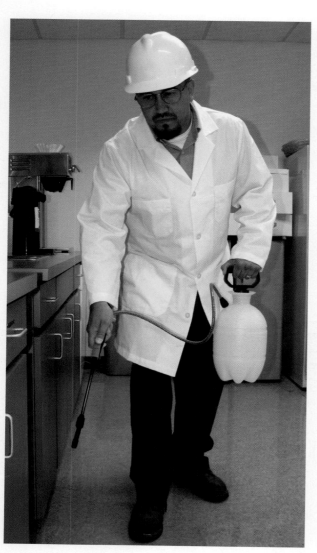

Pest Control Operators (PCOs)

Pest control operators are persons licensed to control pests in the state in which they operate. They take and pass an examination administered by the state to receive their license. In order to maintain their license, they must attend training on a continual basis.

Working as a team

An integrated pest management (IPM) team is made up of food service employees, primarily the establishment manager, kitchen staff and servers. The facility's PCO is also a valued team member. The PCO's role is to assist and support the facility team at each step in the IPM process. The PCO supports the team by helping establishment team members:

- Look for signs of pests. The PCO has spent more time with pests than other team members. This knowledge is helpful in learning about which pest might be in a facility and to find out how they got there.
- Identify pests. If found, the PCO can help the team determine which pests are a serious problem, which are a minor problem, and what action to take and when.
- Determine a control method. Control methods are one of the PCO's greatest contributions to the team. What would control a pest problem the best? Improvement in the cleaning schedule? Physical methods—traps, barriers, sticky pads? Pesticides?
- Evaluate progress. If pests are resistant to a specific control recommended, is there a better control? If not, what are other alternatives to avoid future occurrences?
- Learn pest control practices. Once knowledgeable of a facility, a PCO can help team members decide when to handle a situation themselves or when to call the PCO.

Advantages of using a PCO

Through initial training, continuous learning and on-the-job learning, PCOs have the most effective pest control or pest extermination treatment experience available to a facility. With this background, they can:

- Quickly identify issues. They know when a pest problem exists, how it will progress and the most efficient means to stop it. Once the problem is controlled, they can recommend countermeasures to prevent future infestations.
- More efficiently select treatments. PCOs are up-to-date on pest control products, equipment and techniques. They know how much and what kind of chemicals to use. They also know that pesticides are not the only or, in many cases, the best option. There can be substantial cost savings by minimizing the use of pesticides in a food service facility.

- Safely apply treatments. ONLY a PCO should apply pesticides in the form of sprays, gases, dusts and fumigants. The PCO will assure that no food preparation is in progress and that all food is put away before applying pesticides.
- Save money. The right treatments, at the right time, in the right amount will prevent overuse of pesticides and retreatments.
- Respond quickly. The PCO can address and correct an infestation, while permitting the food service staff to do what it does best–prepare and serve food.
- Provide advice and record keeping. As a member of a facility's pest control team, the PCO can provide good housekeeping know-how, conduct inspections and maintain records of the facility's overall pest control effort.

Selecting a PCO

Selecting a PCO from an ad or a referral is a good way to get started, but it shouldn't be the only approach. Use other resources at your disposal, including other food service providers. When selecting a PCO, do the following:

- Talk to other food service managers in your area. Find out which pest management company or individuals they use and how satisfied they are with their service.
- Deal only with qualified and licensed pest management companies or individuals. Ask which license or licenses they hold, ask to see them and make sure that they are up-to-date. Inquire into a candidate's membership of professional organizations. Check online for membership of the National Pest Management Association.
- Find out if the pest management company has liability insurance to cover any damages to your facility and furnishings. Ask for verification.
- Ask if their work is guaranteed. If so, what is covered and for how long? Thoroughly review any service contracts offered and make sure they meet all needs.
- Take time to make a decision. Seek value, not just cost. Does the person meet all your needs? Is it someone you would trust?

Pesticides

Pesticides are chemicals used to kill pests. By their very definition, they are intended to harm. Misused, pesticides can cause illness and physical conditions in humans far greater than the damage the pests themselves could have inflicted.

Pesticide packaging and distribution

Pesticides are packaged and distributed in many forms. Some examples include:

- Baits, bait traps, glue boards and tapes—a mixture of food to attract and pesticide to kill. Bait stations can permit the insect to ingest the poison and return to their colonies; bait traps, glue boards and tapes are designed to hold the pests. Bait stations are a popular, longer-term solution in eliminating rodents. Glue boards without pesticides are also used for rodent control, although they can be ineffective since rodents can escape.
- Dusts and powders—dry mixtures used to get into the crack and crevices where insects hide.
- Liquids—pesticides in concentrated form that require dilution before use. Some pesticides are distributed as ready-to-use liquids to get into cracks and crevices, like dusts and powders.
- Sprays, including contact, fogger, fumigant, residual.

Contact sprays are multiple-spray, aerosol applicators that are sprayed directly on to pests. Foggers are complete release aerosol applicators used to clear a large area of an infestation. Fumigants are extended use chemical compounds that vaporize over a long period of time. Residual sprays are also extended use pesticides applied to surfaces for a long lasting effect.

Sprays should never be used when food is out in the open.

Do not use pesticides yourself

Pesticides must be treated with respect. Even the most seemingly harmless products can cause damage if used improperly.

The following are possible outcomes from improper use of pesticides by food service workers:

- Unnecessary exposure. Using non-chemical controls is the best protection from the unwanted effects of pesticides. Pesticides should only be used as a last resort. What other control options are available? How effective are they compared to the pesticide?
- Failure to remedy infestation. Has the pest been correctly identified? Was the application too much? Too little? Was there a need for a retreatment? Has the pest built up immunity to the pesticide?
- Introduce a great hazard. Failure to follow label instructions. Was the pesticide for food service use? Was an outdoor-only pesticide used indoors?
- Contaminate food. Was a pesticide applied with food present? Utensils removed and equipment covered? Were food preparation work areas cleaned after application? Was the area properly ventilated?
- Health hazard to a food service worker. Did the label require use of personal protective equipment? Should rubber gloves and eye goggles be worn during application? Should respiratory protection be worn?

Using pesticides

If a situation should arise in which you must use a pesticide **(only a PCO may apply a restricted-use pesticide)**, food service workers should follow these guidelines when selecting and applying pesticides:

- Read and follow ALL label directions. Abuse of pesticides results most commonly from not reading label instructions. For additional information about a pesticide, obtain a Material Safety Data Sheet (MSDS). An MSDS is required for every pesticide and can be found online.
- Use and store pesticides in their original containers—ALWAYS. In addition to guaranteeing proper identification, this will assure that the product label is readily available for future reference or for other users.
- Apply according to label directions. Application directions are established based on the type of pest and threshold levels for effectiveness.
- Wear personal protective equipment. In addition to label information, the MSDS has very specific directions for protecting the user, including what actions to take in case of contact, inhalation or ingestion.

Storing and disposing of pesticides

Improper storage of pesticides can create a health hazard, as well as compromise the effectiveness of the pesticide. Improper disposal can not only pass these same health hazards outside the facility but also harm the environment. The following are guidelines for storing and disposing of pesticides:

- Follow all label and Material Safety Data Sheet (MSDS) directions for storage and disposal. Contact your PCO for assistance in compliance with all pesticide storage and disposal practices. Some state and local governments also have rules that may be stricter.

- Buy only the amount of pesticides needed. Purchase only enough for a single infestation, or for a current season. For disposal, an empty container is preferred to a full container.

- Store within temperature guidelines. Too high or too low temperatures can reduce the effectiveness of a pesticide. Too high temperatures for aerosol cans (typically above 120°F [49°C]) can lead to a release and possible fire hazard.

- Store in a sealed cabinet away from food products, food preparation, and other cleaning materials. Include inspection criteria for cabinets on the cleaning schedule. Consider storage in an outside area, temperature permitting. Label as "Pesticide Storage."

- Never dump unused pesticides into the sewer system. Some water facilities are not equipped to filter out the chemicals. Follow the guidance of your PCO and label directions for safe disposal. Or, offer unused, labeled pesticides to another facility for its use. If the community has a waste disposal day, save them for drop off.

1. A good pest control program will help prevent each of the following except:

   a. Loss of customers and profits caused by selling contaminated food
   b. Loss of customers and profits caused by selling poor quality food
   c. Foodborne illnesses and foodborne disease outbreaks
   d. Contamination and food waste

2. What is integrated pest management (IPM)?

   a. An approach to pest control in which a wide range of practices are used to prevent or solve pest problems
   b. A government agency dedicated to managing pests in all food establishments
   c. A collaboration of various food establishments working together to solve their pest problems
   d. A method of pest control that relies solely on pesticides

3. Which two practices does a successful integrated pest management (IPM) program balance?

   a. Elimination and sanitation
   b. Prevention and control
   c. Removal and prevention
   d. Removal and sanitation

4. What are some ways that food pests can enter a building?

   a. Through gaps or holes in the building
   b. Brought in via food delivery
   c. Brought in via a person
   d. All of the above

5. What are two things you should do when it comes to pest control?

   a. Keep food storage rooms dark
   b. Keep food storage rooms cool
   c. Deny pests an adequate water supply
   d. Deny pests an adequate food supply

6. What is one way customers can help minimize pests at outdoor dining areas?

   a. Do not throw away food wrappers
   b. Do not dine near a dumpster
   c. Do not wear brightly colored clothing
   d. Do not feed wildlife

7. Which pest has the least ability to spread disease?

   a. Cockroaches
   b. Rodents
   c. Ants
   d. Flies

8. Which two of the following are required for licensed PCOs?

   a. They must perform at least two years of apprenticeship training
   b. They must take and pass an exam administered by the state
   c. They must attend training on a continual basis
   d. They must take and pass an annual exam

9. Each of the following is an advantage of using a PCO except:

   a. Faster identification of pest issues
   b. More efficient selection of treatments
   c. Better positioning within the food industry
   d. Safer application of treatments

10. What is the best resource for additional information about a pesticide?

    a. Fellow food industry managers
    b. The FDA *Food Code*
    c. Material Safety Data Sheet (MSDS)
    d. Pesticide Safety Data Sheet (PSDS)

LEGAL REQUIREMENTS, HACCP AND INSPECTIONS

GOVERNMENT FOOD SAFETY REGULATIONS

LESSON 1

After completing this lesson, you should be able to:

- Describe food safety roles and regulations at the federal level
- Describe the purpose and application of the FDA *Food Code*
- Explain how issues and supplements to the *Food Code* are offered
- Explain who develops the *Food Code*
- Describe the topics covered by the *Food Code*

>TERMS TO KNOW

CDC Centers for Disease Control and Prevention. The CDC investigates foodborne disease outbreaks.

EPA Environmental Protection Agency.

FDA Food and Drug Administration.

NMFS National Marine Fisheries Services.

USDA United States Department of Agriculture.

The United States has a long history of government food regulation that involves all levels of government.

Federal Level

While there are no federal food safety regulations that apply to the majority of food establishments, the federal government does play an important role in ensuring the safety of the U.S. food chain from beginning to end.

Several federal agencies work to ensure the safety of food before it ever enters your establishment.

Many of these same agencies are involved in issuing the FDA *Food Code*, the federal government's "best advice" to minimize the incidence of foodborne illnesses.

Most state and local governments have adopted the FDA *Food Code* (or a modified form of it) to regulate the food establishments under their jurisdiction.

Centers for Disease Control and Prevention

The CDC's primary concern in terms of food safety is the prevention and surveillance of foodborne disease outbreaks.

If an outbreak occurs, the **CDC** works with all levels of government to provide a rapid response.

Environmental Protection Agency

The **EPA** determines air- and water-quality standards, regulates pesticide use and dictates proper handling of waste products.

Food and Drug Administration

The **FDA** is responsible for inspecting food service operations that cross state borders. They're also responsible for food products not covered by the **USDA** or the NMFS.

The FDA regulates the medicines and feed consumed by food animals to make sure those animals are safe for people to consume.

The FDA also inspects food-processing plants to ensure compliance with sanitation, labeling and other quality standards.

National Marine Fisheries Services

Much like the USDA, the **NMFS** inspects and grades domestic and imported fish and fish products prior to sale. The NMFS is part of the National Oceanic and Atmospheric Administration (NOAA).

The NMFS also inspects domestic fish-processing operations to ensure compliance with sanitation, labeling and other quality standards.

U.S. Department of Agriculture

Through its Food Safety and Inspections Service (FSIS) subagency, the USDA is responsible for inspecting and grading domestic and imported food and food products derived from domesticated animals like cattle and chickens, including eggs and egg products.

The USDA also inspects domestic slaughterhouses for compliance with sanitation, labeling and other quality standards.

State Level

Safety regulations that affect food establishments are typically written into law at the state level. The only exceptions are international or interstate establishments like those on planes and trains. Such establishments follow federal law and are subject to inspection by federal agencies, more often than not the FDA. Each state decides whether to adopt the FDA *Food Code* in its entirety, in a modified form or not at all. It's the responsibility of state regulatory authorities to ensure that local authorities perform regular inspections and enforce state laws during those inspections. Simply put, some states have stricter or more lenient regulations than those put forth in the federal government's model food code. Further, states don't always revise their own food safety regulations with each update of the FDA *Food Code*. The types of food establishments subject to inspection and the frequency of inspection also differ from state to state.

Food safety regulations aren't just written into law at the state level; they're also enforced at the state level. If your establishment is located in a small city or a rural area, a state (rather than local) regulatory authority may conduct your inspection. In any case, it's the responsibility of state regulatory authorities to ensure that local authorities perform regular inspections and enforce state laws during those inspections.

Local Level

In most instances, local authorities are responsible for enforcing a state's food safety regulations through inspections. City and county health departments conduct the bulk of inspections. As a manager, your most direct contact with government will be at the local level.

Be aware that enforcement systems (the type and number of inspections, and their frequency) can vary by city or county.

If you have questions about the food safety regulations that affect your establishment, it's best to begin by contacting your local health authority.

Your Responsibility

Your responsibilities with respect to government regulations include:

- Familiarizing yourself with the agencies involved in the U.S. food safety system
- Determining which federal, state and local agencies regulate your establishment
- Obtaining copies of any regulations to which your establishment is subject

FDA *FOOD CODE*

As a model food code, the FDA *Food Code* represents the federal government's "best advice" for minimizing the incidence of foodborne illness. It's not a law, however, regulatory authorities at all levels of government typically use the *Food Code* to develop or update their own food safety rules. In fact, according to the Association of Food and Drug Officials, 85% of the U.S. states and territories have implemented some form of the FDA *Food Code* as of June 2005. Another good reason you should be familiar with the *Food Code* is because it has nationwide application. If you transfer to a restaurant or other food establishment in another state, the *Food Code* will still apply even if state and local regulations change.

The FDA issues the *Food Code* every four years and then amends it every two years with the release of a supplement. While the *Food Code* bears FDA's name as author, the FDA isn't the only group involved in creating it. Other federal agencies, mainly the USDA and the CDC, work with the FDA to create the *Food Code*.

The only group outside the government that has much input into the contents of the *Food Code* is the Conference for Food Protection. The Conference for Food Protection, or CFP, is a nonprofit organization formed in 1971 that seeks to create an equitable partnership among food industry regulators, professionals, academics, and consumers. Their conference takes place every two years, before the release of a revision or supplement to the FDA *Food Code*. The goal of the conference is to identify problems, formulate recommendations, and develop practices to ensure food safety. The CFP then presents their findings to the FDA for consideration as they develop new issues or supplements to the FDA *Food Code*.

The current FDA *Food Code* includes hundreds of pages of recommendations that cover the following topics:

- Management and personnel (including supervision, employee health, personal cleanliness and hygienic practices)
- Food preparation and handling (including criteria for receiving, storing, displaying, handling, preparing, serving and transporting)
- Equipment, utensils, and linens (including design, construction, operation, maintenance, repair, cleaning and sanitizing procedures and protection)
- Water, plumbing, and waste (including utilities and services)
- Physical facilities (including design, construction, maintenance, repair, capacities, location and operation)
- Poisonous or toxic materials (including identification and labeling, storage, sale and application)
- Compliance and enforcement (including code applicability, plan submission and approval, permit to operate, inspection and correction of violations)

HAZARD ANALYSIS AND CRITICAL CONTROL POINT

LESSON 2

After completing this lesson, you should be able to:

- Explain the purpose of the HACCP approach
- List the seven principles of the HACCP approach

In 1993 the FDA began including Hazard Analysis and Critical Control Point (known as **HACCP**) principles in the *Food Code*. The hazard analysis and critical control point (HACCP) approach is a food safety management system that focuses on the control of hazards throughout the food flow rather than on sanitation alone. HACCP food safety programs are mandatory for most food manufacturers and also for schools. Some states have mandated HACCP programs at the retail level, and conduct HACCP-based inspections in food establishments. You can expect this trend to continue because many regulators consider the HACCP approach to be the best defense of the food chain "from farm to table."

History

The HACCP approach began with NASA's Apollo space program. A project of the Pillsbury Company, the original goal of an **HACCP plan** was to ensure that astronauts had safe food. The consequences of astronauts suffering from diarrhea or vomiting in a zero-gravity environment would be catastrophic and could lead to mission failure.

Before HACCP programs, most reputable food establishments were aware of potential food safety problems and implemented appropriate controls. But the food industry made very little attempt to determine which controls were most important to reduce the risk of foodborne illness. Instead, food safety control relied heavily on end-product testing, which had many disadvantages:

- Control was reactive, that is, action was taken after the problem
- Testing was often slow and results would take considerable expertise to interpret, which meant that sometimes the product reached consumers before a problem was diagnosed
- The cost of sampling and analysis was high
- Only a few staff members were directly involved in food safety

Purpose and Principles

What's unique about the HACCP approach as a food safety management program is that it's proactive rather than reactive. In other words, a HACCP program tries to prevent foodborne illnesses before they occur instead of dealing with illnesses after the fact.

The HACCP approach examines each step of the food-handling process and identifies potential hazards. The idea is that identifying hazards in advance will help prevent them from ever becoming an end-product problem.

A HACCP program takes time and is not something you should implement quickly or haphazardly for the benefit of an inspector. It requires mapping and examining even the most mundane food-handling processes. But once it's properly implemented, the HACCP approach can help you ensure consistently high standards of food safety, every working day.

The process approach to HACCP can best be described as dividing the food preparation methods in an establishment into categories based on the number of times a specific food will be in the temperature danger zone, then analyzing the hazards and placing managerial controls on each grouping.

PROCESS 1:
Food Preparation with No Cook Step
Example flow: Receive-Store-Prepare-Hold-Serve (Other food flows are included in this process, but there is no cook step to destroy pathogens.)

PROCESS 2:
Preparation for Same Day Service
Example flow: Receive–Store–Prepare–Cook–Hold–Serve (Other food flows are included in this process, but there is only one trip through the temperature danger zone)

PROCESS 3:
Complex Food Preparation
Example flow: Receive-Store-Prepare-Cook-Cool-Reheat-Hot Hold-Serve (Other food flows are included in this process, but there are always two or more complete trips through the temperature danger zone.)

The Process Approach

The process approach to HACCP programs can best be described as dividing the food preparation methods in an establishment into categories based on the number of times a specific food will be in the temperature danger zone, then analyzing the hazards and placing managerial controls on each grouping.

Seven HACCP Principles

1. Conduct a hazard analysis to identify the hazards across the process or operation and specify **control measures**.

2. Determine critical control points (CCPs) to pinpoint which of the steps where hazards were identified are critical to food safety.

3. Establish critical limits, **target levels** and tolerances for each CCP.

4. Establish a monitoring system for each CCP, through scheduled testing or observations.

5. Establish **corrective actions** to be taken when a CCP is out of control, that is, when a critical limit is breached.

6. Establish **verification** procedures, which include appropriate validation, together with a review to confirm that the HACCP program is working effectively.

7. Establish documentation and record keeping for all procedures relevant to these principles and their application.

Conducting a hazard analysis

The goal of this type of analysis is to identify the hazards across a process or operation and to specify control measures.

All other HAACP principles depend on the **hazard analysis**. The choices you make here determine the scope and success of your HACCP program. The hazard analysis consists of six steps.

The first step: Assemble a team of staff members to conduct the analysis

Team members should be familiar with the selected products and processes. It's important for you, as the manager, to not just be part of the team but also to support and lead it. You need to ensure that all team members have the time, resources and training they need to complete the analysis.

The second step: Determine which products and processes you need to examine

In a food service establishment, examples include perishable raw food, which is cooked and served hot, TCS (PHF) food, which is served cold, and frozen raw food, which is cooked, cooled and served cold.

The third step: Prepare flow diagrams of any related processes

A flow diagram is a map of the sequence of steps or operations involved with a particular food item or process, usually from receipt of raw materials to the consumer. Team members then validate the flow diagrams to make sure they accurately reflect what happens in actual practice. It's often beneficial to include a map of the food preparation area layout to help identify potential hazards such as cross-contamination.

The fourth step: Brainstorm potential hazards

Potential food hazards include but are not limited to the following:

- Poor temperature control or prolonged time at ambient temperatures can result in any food poisoning bacteria that are present in food multiplying to large numbers.

- Failure to cook food thoroughly can result in the survival of food poisoning bacteria, which in turn can make anyone who eats the food ill.

- Physical or chemical hazards can occur at any stage in the process. It's unlikely that their removal can be guaranteed later on in the process.

The fifth step: Perform a risk assessment to determine the significance of potential hazards

Whether a hazard is considered significant depends on the likelihood of the hazard occurring—that is, its **risk** and the seriousness of potential consequences. At this point, low-risk hazards that don't have severe consequences are set aside and excluded from the hazard analysis. Such hazards are best addressed as quality control issues.

The final step: Determine control measures for high-risk hazards

Control measures are actions required to prevent or eliminate a food safety hazard or to reduce the hazard to an acceptable level. Controls can be general or specific. General controls relate to prerequisites such as using approved suppliers, effective cleaning and sanitizing, pest management, staff supervision and training, good design and effective maintenance. Specific controls include monitoring items such as the time and temperature of cooking or the volume or weight of preservatives added to a food product.

Determining critical control points

The goal of determining critical control points is to identify which steps in a potentially hazardous process are critical to food safety.

A critical control point (or **CCP**) is a step in the process where it is possible to apply a control that is essential to preventing or eliminating a food safety hazard or reducing it to an acceptable level. Depending on your operation, control measures may be effectively implemented in your prerequisite programs.

Examples include:

- Final cooking temperatures
- Hot-and-cold holding temperatures
- Cooling times
- Food pH and water activity (A_w)

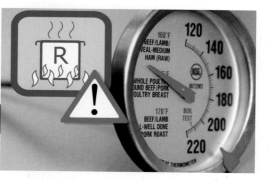

Establishing critical limits

The goal of establishing critical limits is to identify target levels and tolerances for each CCP.

Critical limits are the values of monitored actions at CCPs that separate the acceptable from the unacceptable. Critical limits must be measurable. The results must be obtained at the premises without sending samples away for laboratory analysis. Where possible, critical limits should involve physical characteristics such as temperature, time and pH, along with the weight and size of food.

Target levels are specified values for control measures that eliminate or reduce hazards at CCPs. They act as a buffer zone, providing a tolerance for operation that can prevent a failure to meet critical limits.

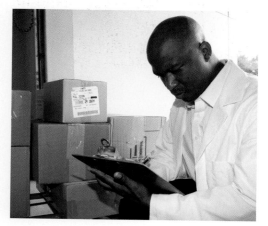

Establishing a monitoring system

The goal of establishing a monitoring system is to create a testing and observation schedule for each CCP.

Monitoring is the planned observation and measurement of control parameters to confirm that the process is under control and that critical limits are not exceeded. Monitoring is required to identify breaches of target levels, to ensure that the process is kept under control, to identify deviations and trigger corrective actions, to provide records for verification and to investigate complaints or for due diligence. Monitoring can be manual or automatic, continuous or at set frequencies. Most importantly, monitoring must permit rapid detection and correction, whatever the type of observation or measurement used.

Observations (smell, touch, sight)

The appearance, smell, texture and other physical characteristics of food are valuable for obtaining a rapid assessment of food standards. For example, food that smells stale or musty indicates a problem. The corrective action would be to discard the food.

Measurements

Measurements include checking temperatures, times and pH values or checking the goods receipt book, cleaning schedules and maintenance records.

Audits

Audits may address the entire premises and procedures or specific practices. They may investigate the cleaning system or the pest management system. Audits of delivery vehicles, drivers and the goods delivered are also needed.

Competency testing

To ensure food safety all the time, all relevant staff must know and understand the importance of CCPs, monitoring, target values and critical limits and the role they play in monitoring procedures. Specific training may be required. You must satisfy yourself that all staff, and not just the HACCP team, are competent in HACCP procedures as needed for their activities.

Whatever methods you use, monitoring systems should state:

- Who is responsible for monitoring and, where necessary, who is responsible for checking that monitoring has been carried out satisfactorily
- What the critical limits, target levels and tolerances are
- When to undertake the monitoring
- Where in the flow diagram monitoring should be done—this should be at the CCP or as close as possible to the CCP
- How to undertake the monitoring, including details of what equipment to use and its calibration

Establishing corrective actions

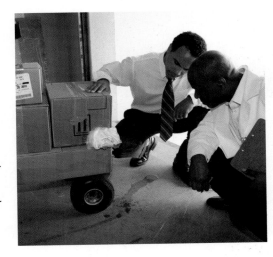

The goal of establishing corrective actions is to identify the procedures to be taken when a CCP is out of control.

Corrective action is the action to be taken when a critical limit is breached.

Corrective action usually has two aspects:

- Dealing with any affected product
- Bringing the CCP and the process back under control

There are several ways products that test outside critical limits can be treated—for example, a hazardous product may be quarantined and subject to further testing, its shelf life may be reduced or the product may be destroyed.

A process that contributed to a hazard may be adjusted and corrected. For example, the cooking time of a product can be extended.

Procedures for corrective action should be documented. They should specify:

- The action to be taken.
- The person responsible for taking action—a clear chain of command is required to avoid delays and ensure that the correct action is taken.
- Who should be notified.
- Who is authorized to stop and then restart production or sales—this will normally be you, the manager.
- The treatment of affected products.

Establishing verification procedures

The goal of establishing verification procedures is to put a system in place to confirm that the HACCP program is working effectively.

Verification involves methods, procedures and tests in addition to those used in monitoring to determine compliance with the HACCP plan.

Verification comprises three stages:

1. Validation
2. Ensuring that the HACCP system is satisfactory
3. Review

Validation involves obtaining evidence that elements of the HACCP plan are effective, especially CCPs and critical limits.

Ensuring that the HACCP system as a whole is satisfactory usually involves auditing against the HACCP plan to ensure correct implementation. You will also need to guarantee that:

- The flow diagram remains valid
- Hazards are being controlled
- Monitoring is satisfactory
- Where necessary, appropriate corrective action has been or will be taken

The final stage of verification is review. The HACCP plan should be reviewed at least annually to ensure that it remains effective. Remember that review documentation must always be recorded.

Establishing a record-keeping system

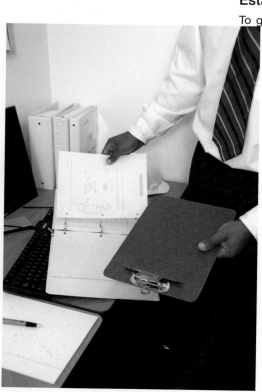

To guarantee the success of any HACCP program, a management system needs to be in place to keep HACCP documentation and records up to date and effective.

You should use the simplest, most effective system that integrates well with your operation. The amount and type of paperwork required to support HACCP systems should be proportionate to the type and size of food business and the risks involved with the processes. Accurate records and documentation are essential for verification and auditing. Records and documentation are also useful for demonstrating how food safety is being managed at your business. It's not uncommon for inspectors to review them. Records are also vital when you need to investigate customer complaints and claims of alleged food poisoning.

The HACCP plan is the main documentation you have.

Your documentation should include the following items:

- Details of the HACCP team and their responsibilities
- Hazard analysis, including product or process descriptions
- Flow diagram and CCP determination
- Critical limits, target values and corrective actions
- Monitoring and record-keeping procedures
- Verification procedures

Other documents that show how you've been supporting your HACCP plan include product specifications, a floor or room plan and details of prerequisite programs such as personal hygiene and pest management.

Examples of useful records are:

- CCP and non-CCP monitoring activities
- Deviations and corrective actions
- Modifications to the HACCP system, including review details
- Audit reports
- Customer complaints
- Calibration of instruments
- Approved supplier list
- Stock rotation records
- Staff health records
- Cleaning schedules and training records

Monitoring records should be signed and dated by the food handler who does the monitoring and countersigned by you or by that person's manager.

Your responsibility

Your responsibilities concerning the HACCP system include:

- Becoming familiar with HACCP principles
- Determining whether your establishment is subject to regulations based on the HACCP principles
- Considering how your establishment would benefit from an HACCP program

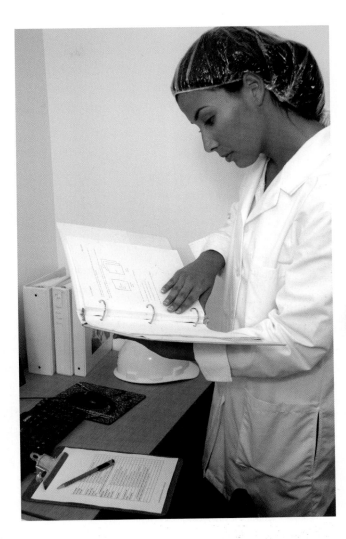

INSPECTIONS

LESSON 3

After completing this lesson, you should be able to:

- Describe the purpose of an inspection program
- Describe the three types of food safety inspections
- Explain how often inspections may be performed
- List some best practices to follow before, during, and after an inspection

>TERMS TO KNOW

CORE ITEM Those hazards that don't have a direct impact on food safety; includes an item that usually relates to general sanitation, operational controls, sanitation standard operating procedures (SSOPs), facilities or structures, equipment design, or general maintenance.

PRIORITY ITEM A provision whose application contributes directly to the elimination, prevention or reduction to an acceptable level of hazards associated with foodborne illness or injury and there is no other provision that more directly controls the hazard; items with a quantifiable measure to show control of hazards such as cooking, reheating, cooling, hand washing.

PRIORITY FOUNDATION ITEM An item that requires the purposeful incorporation of specific actions, equipment or procedures by industry management to attain control of risk factors that contribute to foodborne illness or injury such as personnel training, infrastructure or necessary equipment, HACCP plans, documentation or record keeping, and labeling.

Inspections

The primary purpose of any inspection program is to protect the public's health by requiring establishments to provide food that is safe, unadulterated and honestly presented.

Inspection is the principle tool a regulatory agency has for detecting procedures and practices that might be hazardous and for taking actions to correct deficiencies.

Serving unsafe food can jeopardize the health of your customers and your establishment. Inspections can tell you how well your establishment is doing at following standard food safety practices. An excellent food safety inspection report is one of the most effective forms of advertisement for your establishment, because inspection data are public information. In fact, inspection data are readily available on the Internet in most jurisdictions.

Establishments Subject to Inspection

Any establishment that serves food to the public is subject to inspection by one or more regulatory authorities. This is true even if there's no charge for the food or if consumption of the food takes place off premises.

Inspection Frequency

While inspection frequency varies by jurisdiction, it's safe to assume that your establishment will be inspected at least every six months. Some regulatory authorities inspect all food establishments on the same schedule. Other authorities inspect high-risk food establishments more frequently.

Factors used to determine risk include:

- Establishment size. Larger establishments employ more people and serve more of the public; hence there is an increased risk of foodborne illness.
- Foods served. Extensive handling of raw foods increases the possibility of foodborne illness.
- Susceptibility of clientele. Children, the elderly, and people with suppressed immune systems are more susceptible to foodborne illness.
- Previous compliance history. Establishments with repeat violations are typically considered high-risk until they achieve a consistent record of compliance.

The Inspection Process

Some inspectors provide advance notice of an inspection. This gives you the opportunity to prepare mentally and logistically for an inspection of your establishment. Most inspectors arrive unannounced during the hours of operation and ask for either the manager or the person in charge.

Your establishment may be subject to any one of three types of inspections—traditional, risk-based or HACCP-based. It's not uncommon for regulatory authorities to employ combinations of these three inspection types.

Traditional inspections

A traditional inspection ends with a number or letter grade. Most jurisdictions use a number-based system for scoring and then assign a letter grade for the benefit of the public.

For each item (violation), an inspector may subtract from one to five points:

- One or two points are subtracted for core items (noncritical violations)
- Four to five points are subtracted for priority foundation items (critical violations)

You'll need to address all core items before the next routine inspection.

You'll have a time limit—usually 48 hours or less—to correct priority foundation items. Continued violation may result in fines or closure.

While inspectors subtract points for each broad category of a violation, they don't subtract points for each instance of a violation. That means an establishment with multiple violations in the same category might receive the same total score as an establishment with a single violation in the same category.

Many regulators, including the FDA, find this aspect of traditional inspections problematic.

Risk-based inspections

Because of what it views as limitations of traditional inspection systems, the FDA now recommends the use of risk-based inspections. Risk-based inspections make a distinction between priority items, priority foundation items and core items.

In a risk-based inspection, **priority items** are those known to cause or to contribute to foodborne illnesses. Priority and **priority foundation items** include:

- Improper holding temperatures
- Inadequate cooking
- Contaminated equipment
- Unsafe food sources
- Poor personal hygiene

Core items are those that don't have a direct impact on food safety. These include factors such as general physical facility conditions and general work practices. Instead of deducting one to five points for a violation, the risk-based inspectors assess an establishment as "in" or "out" of compliance for one or more risk factors. The inspector will set a time limit for the correction of all item violations. If you're still in violation of any regulation when the time limit has expired, your food establishment may be subject to fines or closure.

HACCP-based inspections

HACCP-based inspections focus on the control of hazards throughout the food flow. During this complex type of inspection, inspectors observe and assess how an establishment receives, stores, prepares and serves food.

HACCP-based inspections don't use a traditional scoring system. As with risk-based inspections, the main goal of a HACCP-based inspection is to determine and resolve priority and priority foundation items.

Best Practices Before an Inspection

Assign a leader

The inspector will ask to speak with the manager or person in charge. You don't want your staff bumbling around trying to figure out who's in charge if you're absent on the day of inspection, so assign someone to take responsibility in your absence.

Have documentation

It's common for inspectors to request records concerning food purchases, pest control or chemical use. Documentation provided during an inspection may become public. Before you're inspected, determine which documents are appropriate to provide and which documents are not. Review your company's policy regarding confidential information or consult with an attorney if you have questions.

Do a test run

Obtain a copy of your regulatory authority's inspection form and do a test run of a typical inspection. You and your staff can take turns playing the role of inspector. Review the results of the test run as a group and make changes as needed before the real inspector arrives.

Perform daily checks

A daily inspection can be a powerful tool. Be sure that:

- Premises and equipment are clean and in good repair and soap, paper towels and cleaning materials are available
- Your staff follows hygiene rules, and their protective clothing is clean
- Signage is satisfactory for your staff and the public
- Food is in good condition and within shelf life, and food and equipment temperatures are satisfactory
- There's no potential for cross-contamination, and there are no traces of pests or other hazards
- All controls are in place and all records are complete, and there's no evidence of staff taking shortcuts

Best Practices During an Inspection

Don't panic

Think of the inspection as a learning opportunity that will benefit your establishment and your customers.

Check credentials

Most inspectors volunteer their credentials immediately on arrival, but criminals may try to gain entry into your establishment by posing as inspectors. If you have doubts about an inspector's credentials, contact their supervisor for verification.

Clarify the reason

It's completely reasonable to ask inspectors whether they're conducting a routine inspection or if they've come because of a customer complaint. Knowing why they've come will help you facilitate the inspection.

Don't refuse

Most inspectors can and will obtain an inspection warrant that you can't refuse. It simply doesn't make good sense to refuse entry to an inspector.

Be cooperative

Being defensive or uncooperative during an inspection may cause the inspector to think you have something to hide. Answer any and all questions the inspector asks as best you can, and require your staff to do the same.

Go along

Accompanying the inspector as they look around will show that you're interested in what they have to say. It also gives you the opportunity to correct minor violations on the spot.

Take notes

As the inspector makes suggestions or points out violations, write them down. While the inspector will provide a written report after the inspection, your own notes will come in handy as you work to improve the safety of your establishment. Taking notes also shows the inspector that you take the visit seriously.

Don't argue

If you disagree with the inspector, don't start an argument during the inspection. Keep quiet for the time being and appeal the decision later. Your goal during the inspection is to make it clear that you're willing to correct any violations.

Act professionally

Avoid offering the inspector any food or other item that could be construed as an attempt to influence the inspection.

Discuss violations

Even the most reputable food establishments sometimes have violations. Don't be confrontational if you're cited. Instead, ask the inspector to discuss the violations and the time frame for correction. The discussion should involve both you and your staff.

Sign the report

Signing the inspector's report doesn't mean that you agree to the findings. It means only that you received a copy of the report.

Best Practices After an Inspection

Act quickly

You are required by law to correct all items (violations) outlined by the inspector. Typically, you'll have less than 48 hours to correct priority and priority foundation items. You must correct any core items as quickly as possible.

Implement fixes

Once your establishment is compliant, it's a best practice to determine why and how any violations occurred. Evaluate your operating procedures and your training procedures to look for flaws. Revise your existing procedures or implement new ones to eliminate problems.

Dealing with suspensions and closures

If an inspector determines that your establishment poses an immediate and substantial threat to the public health, he or she can suspend your permit to operate or request a voluntary closure. In either instance, your establishment must immediately cease operation and eliminate any hazards. Such hazards typically include significant sewer, water or electrical issues, fire or flood, insect or rodent infestation or outbreaks of foodborne illness. Before you can resume operation, your establishment must successfully pass one or more inspections.

Suspensions versus voluntary closures

The end result of either a mandatory suspension or voluntary closure is the same—your establishment must cease operation. However, there are some important differences between the two actions.

Suspensions can be imposed by an inspector, but they require the approval of the local health department. This means your establishment can request a hearing to dispute the suspension. A public notice is usually posted on the front door of an establishment when its permit to operate is suspended.

An inspector can usually request a voluntary closure without the approval of anyone else. The inspector is essentially asking your establishment to close until you've corrected priority and priority foundation food safety issues. Voluntary closures don't typically require public notice.

ASSESSMENT

1. Which of these is the federal government's "best advice" to minimize the incidence of foodborne illness?

    a. USDA Food and Safety Regulations
    b. EPA Food Policies
    c. CDC Criteria on Food
    d. FDA *Food Code*

2. Which agency is responsible for inspecting and grading domestic and imported food and food products derived from domesticated animals like cattle and chicken?

    a. FDA
    b. CDC
    c. USDA
    d. EPA

3. Which is NOT your responsibility with respect to government food safety regulation?

    a. Finding out which inspectors are more lenient and dealing only with them
    b. Becoming familiar with the agencies involved in the U.S. food safety system
    c. Determining which federal, state and local agencies regulate your establishment
    d. Obtaining copies of any regulations to which your establishment is subject

4. What does a flow diagram provide for hazard analysis?

    a. A map of the sequence of steps or operations involved with a particular food item or process
    b. The actions required to prevent or eliminate a food safety hazard
    c. A step where it is possible to apply a control that prevents or eliminates a food safety hazard
    d. Specified values for the control measure that eliminates or reduces hazards at CCPs

5. What term describes specified values for control measures that eliminate or reduce hazards at CCPs?

    a. Control measures
    b. Flow diagrams
    c. Target levels
    d. HACCP principles

6. Which monitoring method would address staff understanding of the importance of CCPs, target values and critical limits?

    a. Verification
    b. Competency testing
    c. Smell, touch, and sight
    d. Audits

7. Which is the correct way to handle a product that falls outside critical limits (e.g., chicken held at 130°F [54°C])?

    a. Destroy the product
    b. Get the product back to the critical limit value
    c. Quarantine the product and subject it to further testing
    d. Serve the product immediately

8. What is the goal of establishing verification procedures?

    a. To create a testing and observation schedule for each CCP
    b. To create a system that confirms that HACCP is working
    c. To identify target levels and tolerances for each CCP
    d. To identify hazards across a process and specify control measures

9. What does an HACCP-based inspection focus on?

    a. The control of hazards throughout the food flow
    b. Violations of good retail practice
    c. A number-based system for scoring
    d. General personal hygiene

10. Which of these is NOT a best practice you should follow during an inspection?

    a. Accompany the inspector throughout the inspection
    b. Make sure the inspector knows when you disagree
    c. Take notes
    d. Act professionally

MANAGER RESPONSIBILITIES

As the manager of a food establishment, you have many responsibilities.

Your responsibilities concerning food safety regulations include:

- Familiarizing yourself with the agencies involved in the food safety system
- Determining which government agencies regulate your establishment
- Obtaining copies of any regulations to which your establishment is subject

Your responsibilities concerning the state or federal food code include:

- Obtaining the most recent issue of your state food code that is based on the FDA *Food Code*
- Becoming familiar with all topics covered in your state food code

Your responsibilities concerning HACCP programs include:

- Becoming familiar with HACCP principles
- Determining whether your establishment is subject to regulations based on the HACCP principles
- Considering how your establishment can benefit from an HACCP program

Your responsibilities concerning inspections include:

- Finding out how to access food safety data for your jurisdiction
- Obtaining copies of your establishment's previous inspection reports
- Determining how often your establishment is subject to inspection
- Becoming familiar with your regulatory authority's inspection process
- Finding out what type of inspection your establishment is subject to
- Training a staff member to act in your absence during an inspection
- Obtaining a copy of your regulatory authority's inspection form
- Implementing a self-inspection system for your establishment

Helpful Resources

Federal Agency Websites:

Centers for Disease Control and Prevention (CDC)
http://www.cdc.gov/

Environmental Protection Agency (EPA)
http://www.epa.gov/

Food and Drug Administration (FDA)
http://www.fda.gov/

National Marine Fisheries Services (NMFS)
http://www.nmfs.gov/

U.S. Department of Agriculture (USDA)
http://www.usda.gov/

CHAPTER ASSESSMENT ANSWERS

CHAPTER 1:

(1) b (2) a,c (3) c (4) d (5) a (6) a,b,e (7) d (8) True (9) c (10) b

CHAPTER 2:

(1) c (2) d (3) a,c (4) d (5) b,e (6) c (7) b (8) e (9) d (10) a,b

CHAPTER 3:

(1) d (2) a (3) b,c (4) c,d (5) b (6) b (7) d (8) a (9) c (10) c

CHAPTER 4:

(1) c (2) a (3) c (4) b,d (5) c (6) b (7) c (8) a (9) a (10) d

CHAPTER 5:

(1) d (2) b,c (3) b (4) a (5) d (6) d (7) b (8) c (9) a,b (10) a,d

CHAPTER 6:

(1) c (2) c (3) b (4) a (5) d (6) b (7) c (8) d (9) a (10) d

CHAPTER 7:

(1) b (2) d (3) a (4) c (5) a (6) b (7) c,d (8) c (9) c (10) a

CHAPTER 8:

(1) b (2) a (3) b (4) d (5) c,d (6) d (7) c (8) b,c (9) c (10) c

CHAPTER 9:

(1) d (2) c (3) a (4) a (5) c (6) b (7) a (8) b (9) a (10) b

THERMOMETER CALIBRATION I

Ice Water Method

To adjust or calibrate the accuracy of an adjustable thermometer using the ice water method, take the following steps:

You will need:

- A bi-metal or digital thermometer (one that can be calibrated)
- A 2 quart (1.89 liters) sauce pan at least 4 inches (10.16 cm) deep
- Crushed ice
- Drinking water
- Adjustment tool or wrench provided with thermometer (this is not needed for thermometers with a reset button)

Directions:

1. Fill the saucepan with crushed ice.
2. Add the water to make a slurry.
3. Stir well and let stand for a minute or two.
4. Insert the thermometer into the ice.
5. Place at least 2 inches (5.08 cm) from the side and bottom of the container.
6. Keep minimum immersion point (circle/dimple on shaft) below water level.
7. Hold in ice water for 30 seconds or until dial stops moving.
8. Place the adjustment tool or wrench on the adjustment nut.
9. Turn the dial to read to 32°F (0°C) while in the ice and tighten adjustment nut. For thermometers with a reset button, push reset while the thermometer is still submerged.

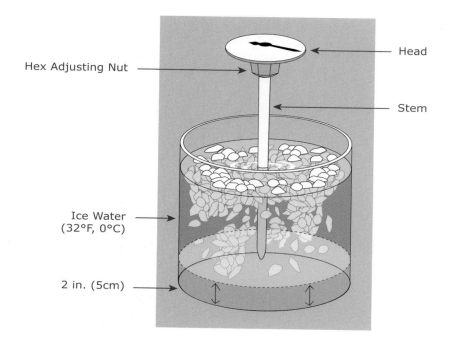

THERMOMETER CALIBRATION II

Boiling Water Method

To adjust or calibrate the accuracy of an adjustable thermometer using the boiling water method, take the following steps:

You will need:

- A bi-metal or digital thermometer (one that can be calibrated)
- A 2 quart (1.89 liters) sauce pan at least 4 inches (10.16 cm) deep
- Heat source stove
- Drinking water
- Pot holders
- Adjustment tool or wrench provided with thermometer (this is not needed for thermometers with a reset button)

Directions:

1. Fill the saucepan with water.
2. Bring to a boil.
3. Remove from heat and place saucepan on pot holder.
4. Insert the thermometer into the boiling water.
5. Place at least 2 inches (5.08 cm) from the side and bottom of the container.
6. Keep minimum immersion point (circle/dimple on shaft) below water level.
7. Wait 30 seconds or until dial stops moving.
8. Place the adjustment tool or wrench on the adjustment nut.
9. Turn the dial to read to 212°F (100°C) while in the boiling water and tighten adjustment nut. For thermometers with a reset button, push reset while the thermometer is still submerged.

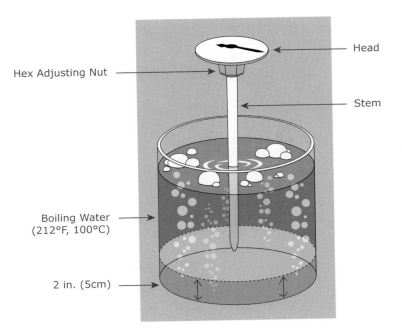

Head

Hex Adjusting Nut

Stem

Boiling Water (212°F, 100°C)

2 in. (5cm)

(The boiling water method of calibration is ineffective in higher altitudes as water boils at a temperature lower than 212°F [100°C].)

GLOSSARY

ACIDIC Having a pH level less than 7.0, as do foods such as vinegar and tomatoes.

ACUTE ILLNESS An illness that develops rapidly and produces symptoms quickly after infection.

ADA The Americans with Disabilities Act. The law requires reasonable accommodation for access to an establishment by clients and employees with disabilities. The law was passed in 1990.

ADDITIVES Preservatives, antioxidants, colorings, emulsifiers, stabilizers, artificial sweeteners and flavorings added to food to improve quality, taste, shelf life, function or appearance.

AEROBE An organism that requires oxygen to live. Also referred to as an "aerobic organism."

AFLATOXINS Toxins produced by mold that can cause serious illness and cannot be killed by cooking.

AIR GAP The space between a water outlet and the highest level of water in a sink drain or tub. It is one of the cheapest and most reliable methods of backflow prevention.

ALKALINE Having the opposite chemical property from acids. Alkaline products have a pH level greater than 7.0.

ALLERGEN Any substance, such as foods, pollens and microorganisms that causes allergies in certain individuals.

ANAEROBE An organism that requires the absence of oxygen to live. Also referred to as an "anaerobic organism."

ANAPHYLACTIC REACTION A severe allergic reaction affecting the whole body, often within minutes of eating the food, which may result in death.

ANISAKIS SIMPLEX A parasite that lives only in its host but survives in food, such as raw or undercooked seafood.

ANTISEPTIC A substance that prevents the growth of bacteria and molds, specifically on or in the human body.

ASEPTIC Free from microorganisms.

BACILLUS CEREUS An intoxication-causing bacteria commonly found in starchy foods and meat products. This type of bacteria produces two types of toxins: emetic and diarrheal. Each toxin causes a different type of illness.

BACKFLOW The reverse flow of water from a contaminated source to the drinking (potable) water supply. It can occur when there is a drop in water pressure and water from the contaminated supply is sucked into the drinking (potable) water system.

BACTERIA Single-celled microorganisms with rigid cell walls that multiply by dividing into two (that is, by binary fission). Some bacteria cause illness and others cause food spoilage.

BIOLOGICAL CONTAMINATION The contamination of food by microorganisms. Examples of biological contaminants include bacteria, parasites, viruses and fungi. These microorganisms may be transferred to food from a variety of sources, such as people, raw food, pests and refuse.

BIOLOGICAL Of or relating to living organisms.

CAMPYLOBACTER JEJUNI An infection causing bacteria found on raw poultry and in contaminated water.

CARRIER A person who harbors, and may transmit, pathogenic organisms with or without showing signs of illness.

CCP Critical control point. A step in the HACCP process where control can be applied and is essential to prevent or eliminate a food safety hazard or reduce it to an acceptable level.

CDC Centers for Disease Control and Prevention. The CDC investigates foodborne disease outbreaks.

CHEMICAL CONTAMINATION The contamination of food by chemical substances such as pesticides and cleaning solutions.

CIGUATOXIN A toxin found in some tropical coral reef fish. The toxin causes the following symptoms when consumed: nausea, vomiting, diarrhea, muscular weakness, numbness in extremities, and it can cause respiratory arrest.

CIP Stands for cleaning place. The cleaning process necessary when equipment cannot be dismantled or moved. It involves the circulating of nonfoaming detergents and disinfectants, or sanitizers, through assembled equipment and pipes, using heat and mostly turbulence to attain a satisfactory result.

CLEANING The process of removing soil, food residues, dirt, grease and other objectionable matter; the chemical used to do this is called a detergent.

CLEANING AGENT A chemical compound, such as soap, that is used to remove dirt, food, stains or other deposits from surfaces.

CLOSTRIDIUM BOTULINUM An intoxication-causing bacteria commonly found in soil and therefore in products that come from soil such as root vegetables. It is anaerobic, which means it grows without oxygen. Because there's no need for oxygen, *Clostridium botulinum* can also be found in improperly canned food.

CLOSTRIDIUM PERFRINGENS A bacteria that causes mild infection from toxin-producing spores. It is anaerobic and can be found in soil, animal and human waste, dust, insects, and raw meat.

COLITIS An acute or chronic inflammation of the large intestine or bowel.

COLONY A group of microorganisms produced from one parent cell.

COMMUNICABLE Capable of being transmitted between persons or species.

COMMINUTED Reduced in size by grinding, mincing, chopping or flaking. Ground meats are examples of comminuted food.

CONTAMINATION The occurrence of any objectionable matter in food or the food environment.

CONTROL MEASURES Actions required to prevent or eliminate a food safety hazard or reduce it to an acceptable level.

CORE ITEM Those hazards that don't have a direct impact on food safety; includes an item that usually relates to general sanitation, operational controls, sanitation standard operating procedures (SSOPs), facilities or structures, equipment design, or general maintenance.

CORRECTIVE ACTION The action to be taken when a critical limit is breached.

CRITICAL LIMIT The value of a monitored action that separates the acceptable from the unacceptable.

CROSS-CONNECTION The mixing of drinking (potable) and contaminated water in plumbing lines.

CROSS-CONTAMINATION The transfer of bacteria from contaminated foods (usually raw) to ready-to-eat foods by direct contact, drip or indirect contact or using a vehicle such as the hands or a cloth.

CRYPTOSPORIDIUM PARVUM A parasite found in soil, food, water or surfaces that have been contaminated with infected human or animal feces.

CYCLOSPORA CAYETANENSIS A microscopic, single-cell parasite that infects the small intestine and is found in contaminated water and various types of produce.

DECLINE PHASE Bacterial growth phase in which the number of bacteria decreases because more bacteria are dying than are multiplying.

DECOMPOSITION The process of decay in a living organism.

DEHYDRATION The process of reducing the amount of water available in food to prevent the growth of microorganisms.

DETERGENT A chemical or mixture of chemicals made of soap or synthetic substitutes. It facilitates the removal of grease and food particles from dishes and utensils and promotes cleanliness, so that all surfaces are readily accessible to the action of disinfectants or sanitizers.

DIARRHEAL Causing diarrhea.

DISINFECTANT A chemical used for disinfection—that is, to reduce microorganisms to a safe level.

DISINFECTION The process of reducing microorganisms to a level that will not lead to harmful contamination or spoilage of food; the chemical used is called a disinfectant. Chemical agents or physical methods used in this process should not adversely affect the food.

E. COLI A bacteria found in the intestines of mammals. It can be found in ground beef and contaminated produce.

EMETIC Causing vomiting.

ENDOTOXIN Toxin (poison) present in the cell wall of many bacteria that is released on death of the bacteria.

ENZYMES Proteins produced by living organisms.

EPA Environmental Protection Agency.

EXCLUSION Requiring a worker to leave the food establishment as a result of specific illnesses, symptoms or exposure to certain diseases.

EXOTOXIN Toxin (poison) usually produced during the multiplication of some bacteria. They are highly toxic proteins and are often produced in food.

FACULTATIVE ANAEROBE An organism that can multiply with or without the presence of oxygen.

FAT TOM The acronym that lists the conditions that support the rapid growth of microorganisms. These conditions are food, acidity, temperature, time, oxygen and moisture.

FDA Food and Drug Administration.

FIFO The acronym for first in first out, which is a system used in stock rotation.

FOOD ADDITIVES Preservatives, food colorings and flavorings that are added to food.

FOOD ALLERGY An identifiable immunological response to food or food additives, which may involve the respiratory system, the gastrointestinal tract, the skin or the central nervous system. The most common food allergies are caused by: nuts (especially peanuts), eggs, wheat, shellfish, milk and soy.

FOOD ESTABLISHMENT Any business (whether for profit or not) whose commercial operations deal with food or food sources.

FOOD IRRADIATION The process of subjecting food to doses of ionizing radiation (such as gamma rays, x-rays or electrons) to destroy parasites, insects and most microorganisms, to extend the life of food and reduce the risk of foodborne illness.

FOOD SAFETY The measures and conditions necessary to control hazards and to ensure fitness for human consumption of a foodstuff, taking into account its intended use.

FOOD SAFETY HAZARD A biological, chemical or physical agent in food, or a condition of food, with the potential to cause harm (that is, an adverse health effect) to the consumer. Most biological hazards are microbiological.

FOOD SAFETY MANAGEMENT SYSTEM The policies, procedures, practices, controls and documentation that ensure that food sold by a food business is safe to eat and free from contaminants.

FOOD SAFETY POLICY A company's commitment to producing safe food, providing satisfactory premises and equipment and ensuring that legal responsibilities are met and appropriate records maintained. The document outlines management responsibilities and is used to communicate standards to staff.

FOODBORNE DISEASE OUTBREAK The same illness in two or more individuals resulting from the consumption of the same food.

FOODBORNE ILLNESS An acute illness resulting from eating contaminated food with symptoms including abdominal pain, diarrhea, vomiting and nausea.

FUMIGATION A method of pest control that completely fills an area with smoke, gas or vapor in order to kill vermin or insects.

FUNGI Biological contaminants that can be found naturally in air, plants, soil and water. Fungi can be small, single-celled organisms or larger multi-cellular organisms.

GASTROINTESTINAL Relating to the stomach or intestines.

GENERATION TIME The time between each bacterial division.

GERMINATION The development or growth of microorganisms.

GIARDIA DUODENALIS A parasite found in contaminated water, raw fruits and vegetables.

HACCP Acronym for hazard analysis and critical control point. A food safety management system that identifies, evaluates and controls hazards that are significant for food safety.

HACCP PLAN   Documentation completed during the HACCP study and implementation. It includes the hazard analysis, the flow diagram, the HACCP control charts, monitoring records, verification details and modifications to the system.

HAND WASHING   The process of cleansing the hands with soap and water to thoroughly remove soil and/or microorganisms to safe levels. Food workers must clean their hands up to their elbows.

HAZARD ANALYSIS   The process of collecting and evaluating information on hazards and on conditions leading to their presence to decide which are significant for food safety and therefore should be addressed in the HACCP plan.

HEAT STABLE TOXIN   A toxin that is not destroyed by cooking food at normal times and temperatures.

HEPATITIS A VIRUS   A disease primarily found in the feces of infected persons. It is spread by infected food workers to ready-to-eat food including deli meats. It can also be spread to produce and salads and can be found in raw shellfish.

HIGH-RISK FOOD   See time/temperature control for safety (potentially hazardous) food.

HIGH-RISK POPULATION   People who are at high risk for foodborne illness, including the elderly, the very young, people who are immunocompromised, pregnant women and allergen-sensitive people.

HISTAMINE   A toxin formed in certain scombroid fish, such as mackerel, mahi mahi and tuna, when they are temperature abused.

HISTIDINE   A natural amino acid present in protein material found in certain scombroid fish, such as mackerel, mahi mahi and tuna. When these fish are temperature abused, it can lead to histamine production and scombroid poisoning.

HIV VIRUS   A retrovirus spread through blood and bodily fluids. The CDC has found no evidence that the HIV virus can be transmitted through food.

HOT HOLDING   The storage of cooked food at 135°F (57°C) or higher, while awaiting consumption by customers.

IMMUNOCOMPROMISED   Having an immune system that is impaired, including the very old, very young and those with a disease or ongoing treatment that weakens the immune system.

INCUBATION PERIOD   The period between infection and the first signs of illness.

INFECTION   A disease caused by the release of endotoxins in the intestine of the affected person. Illnesses caused by infection will normally have a longer onset time. It may take one or two days before the infection makes a person feel ill.

INTOXICATION   An illness caused when bacteria produce and release exotoxins into food. Illnesses caused by intoxication will normally have a short onset time. Intoxication can also be caused by chemical residues and food additives.

JAUNDICE   A yellowish discoloration of the skin and eyes, indicating liver malfunction and illness.

LAG PHASE   Bacterial growth phase when bacteria are not multiplying at all.

LESION   A skin injury usually caused by disease or trauma.

LISTERIA MONOCYTOGENES   An infection causing bacteria naturally found in soil, raw vegetables and milk that has not been properly pasteurized. It is associated with certain ready-to-eat foods such as deli meats and hot dogs.

LOGARITHMIC PHASE   Bacterial growth phase when bacteria multiply rapidly.

LOW-RISK FOOD   An ambient stable food that does not normally support the multiplication of pathogens. This category includes foods that are acidic, high in sugar or high in salt.

MAP   Acronym for modified atmosphere packaging. A physical preservation method that involves changing the proportion of gases normally present around a food item; for example, vacuum packing.

MICROBIOLOGICAL   Of the branch of biology dealing with the structure, function, uses and modes of existence of microorganisms.

MICROORGANISMS   Organisms such as bacteria, viruses, fungi and parasites that are too small to be seen with the naked eye. The microorganisms listed contaminate food and cause foodborne illness.

MOLD   Microscopic chlorophyll-free fungi that produce thread-like filaments; they can be black, white or of various colors.

MONITORING   The planned observations and measurements of control parameters to confirm that the process is under control and that critical limits are not exceeded.

NMFS   National Marine Fisheries Services.

NON-PERISHABLE FOOD   Food that does not sustain the growth of microorganisms, such as sugar, flour and dried fruit. Also referred to as "stable."

NOROVIRUS   Also called viral gastroenteritis, is found in the feces of infected persons. Can also be found in contaminated water.

ONSET PERIOD   The period between eating contaminated food and the first signs of illness.

PARASITE   An organism that lives and feeds in or on another living creature, known as a host, in a way that benefits the parasite and disadvantages the host. In some cases, the host eventually dies.

PASTEURIZATION   A heat treatment of food at a relatively low temperature that destroys the vegetative pathogens and most spoilage organisms to safe levels, thus prolonging the shelf life. Toxins and spores generally survive and rapid cooling and refrigerated storage is usually essential.

PATHOGEN   Disease-producing organism.

PERISHABLE FOOD   Food that is most prone to spoilage, such as meat, poultry, fish, dairy products, fruits and vegetables.

PERSON IN CHARGE   An individual responsible for a food establishment, or the department or area of a food establishment.

PERSONAL HYGIENE   Standards of personal cleanliness habits, including keeping hands, hair and body clean and wearing clean clothing in the food establishment.

PEST   An animal, bird or insect capable of directly or indirectly contaminating food.

pH   A index used as a measure of acidity/alkalinity, on a scale of 1 to 14. Acidic foods have pH values below 7 and alkaline foods above 7; a pH value of 7 is neutral.

PHYSICAL CONTAMINATION Occurs when any foreign object becomes mixed with food and presents a hazard or nuisance to those consuming it.

POTABLE WATER Water that is safe to drink; an approved water supply.

POTENTIALLY HAZARDOUS FOOD See time/temperature control for safety food.

PRESERVATION The treatment of food to prevent or delay food spoilage and inhibit the growth of pathogenic microorganisms, which would render the food unfit for consumption.

PRIORITY ITEM A provision whose application contributes directly to the elimination, prevention or reduction to an acceptable level hazards associated with foodborne illness or injury and there is no other provision that more directly controls the hazard; items with a quantifiable measure to show control of hazards such as cooking, reheating, cooling, hand washing.

PRIORITY FOUNDATION ITEM An item that requires the purposeful incorporation of specific actions, equipment or procedures by industry management to attain control of risk factors that contribute to foodborne illness or injury such as personnel training, infrastructure or necessary equipment, HACCP plans, documentation or record keeping, and labeling.

PROTOZOA Single-celled microscopic organisms (larger than bacteria) that form the basis of the food chain. They live in moist habitats such as oceans, rivers, soil and decaying matter; some are pathogenic.

QUALITY ASSURANCE A planned and systematic approach to ensuring that a product or service meets the required standards over time. It is a proactive process.

QUALITY CONTROL A series of techniques used to assess compliance with a standard specification, which relies on end-product testing as an indicator of consistent quality. It is a reactive process.

QUATERNARY AMMONIUM COMPOUNDS A common example of a chemical sanitizer also referred to as quats.

RAPID COOLING The process of cooling food quickly to 41°F (5°C).

READY-TO-EAT FOOD Food that is meant for consumption without any treatment that is intended to destroy any pathogens that may be present. They include all high-risk foods and such foods as fruit, salad, vegetables and bread.

REHEATING The process of recooking previously cooked and cooled foods to a temperature of at least 165°F (74°C).

RESTRICTION Preventing a worker with certain illnesses or symptoms from working with food or in food contact areas.

RISK The likelihood of a hazard occurring in food.

RISK ASSESSMENT The process of identifying hazards, assessing risks and severity and evaluating their significance.

SAFE FOOD Food that is free of contaminants.

SALMONELLA SPP. An infection-causing bacteria commonly found in raw poultry, eggs, raw meat and dairy products. It has also been found in ready-to-eat food that has come into contact with infected animals or their waste.

SANITIZE To use chemicals or heat to reduce the number of micro organisms to a safe level. It is important to note that disinfectants and sanitizers are not the same thing. Disinfectants are not permitted in food preparation areas.

SCOMBRID A member of a family of fish, including tuna, mahi mahi and mackerel, that can produce histamine toxins when temperature abused.

SEPTIC Infected by bacteria, which produce pus.

SHIGA TOXIN-PRODUCING _E. COLI_ An infection-causing bacteria found in ground beef and contaminated produce. It takes only a small amount of the bacteria to cause illness.

SHIGELLA SPP. A bacteria found in the feces of people with Shigellosis. It can be found in ready-to-eat foods such as greens, milk products and vegetables and also in contaminated water. The most common method of transmission is cross-contamination. Flies can also be carriers of this type of bacteria.

SLACKING The process of gradually increasing frozen food from a temperature of –10°F to 25°F (–23°C to –4°C) to facilitate even heat distribution during the cooking process.

SPORE A resistant resting phase of bacteria, protecting them against adverse conditions such as high temperatures.

STAPHYLOCOCCUS AUREUS An intoxication-causing bacteria commonly found on the skin, nose and hands of one out of two people. It is transferred easily from humans to food when people carrying the bacteria handle the food without washing their hands. This bacteria also produces toxins that multiply rapidly in room-temperature food.

STATIONARY PHASE Bacterial growth phase when the number of bacteria produced by multiplication equals the number of bacteria dying.

STOCK ROTATION The practice of ensuring the oldest stock is used first and that all stock is used within its shelf life.

SULFITE PRESERVATIVE A chemical preservative used to prevent spoilage. Sulfites can cause reactions, including respiratory complications, in some individuals.

TARGET LEVEL The predetermined value for the control measure that will eliminate or control the hazard at a control point.

TEMPERATURE DANGER ZONE The temperature range at which most foodborne microorganisms rapidly grow. The temperature danger zone is 41°F to 135°F (5° to 57°C).

TIME/TEMPERATURE CONTROL FOR SAFETY FOODS (TCS) Products that under the right circumstances support the growth of foodborne illness causing microorganisms.

TOXIN Poison produced by pathogens, either in the food or in the body, after consumption of contaminated food.

TRANSMISSIBLE Capable of being transmitted from person to person.

TRICHINELLA SPIRALIS An intestinal roundworm that is found in wild game animals and in undercooked pork. The larvae of the _Trichinella spiralis_ can move throughout the body, infecting various muscles and causing the infection Trichinosis.

TRICHINOSIS An infection caused by _Trichinella spiralis_, which is an intestinal roundworm that can be found in wild game animals and in undercooked pork. The larvae of the

Trichinella spiralis can move throughout the body, infecting various muscles.

UHT Ultra heat treatment. A high-temperature preservation method used to extend shelf life without the changes in flavor and texture caused by sterilization.

UNADULTERATED Food in a pure state.

USDA United States Department of Agriculture.

USE-BY DATE The last date recommended for the use of the product if it is to be used at peak quality. The manufacturer of the product usually determines the date.

VEGETATIVE Capable of growing. Bacteria in the vegetative state continue to divide at regular intervals while conditions are suitable for growth and multiplication.

VEHICLES The objects by which viruses are transferred from sources to ready-to-eat foods, including hands, cloths and equipment, hand-contact surfaces and food-contact surfaces.

VERIFICATION The application of methods, procedures and tests to determine compliance with the HACCP plan, in addition to the monitoring.

VIBRIO PARAHAEMOLYTICUS An infection causing bacteria commonly associated with raw or partially cooked oysters.

VIRUSES Microscopic pathogens (smaller than bacteria) that multiply in the living cells of their host.

WAREWASHING Automatic washing machines or equipment.

WATER ACTIVITY A measure of the water in food available to microorganisms; it is represented by the symbol A_w. Most bacteria multiply best in food with a water activity of between 0.95 and 0.99.

YEASTS Single-celled microscopic fungi that reproduce by budding and grow rapidly on certain foodstuffs, especially those containing sugar.

INDEX

PHOTO CREDITS

CONTENTS

III (Top): © Monkey Business Images; III (Center): © Sebastian Kaulitzki, 2012. Used under license from Shutterstock.com; IV (Top): Image Copyright Peter G, 2012. Used under license from Shutterstock.com; IV (Bottom): Image Copyright Fedor Kondratenko, 2012. Used under license from Shutterstock.com; V (Top): © MindLeaders; V (Center): © Anson0618 , 2012. Used under license from Shutterstock.com; V (Bottom): © michaeljung, 2012. Used under license from Shutterstock.com.

CHAPTER 1

Chapter opener: © Monkey Business Images ; Page 1: © Efired, 2012. Used under license from Shutterstock.com.; Page 1: © Stephanie Frey, 2012. Used under license from Shutterstock.com; Page 2: © MindLeaders; Page 3: © holbox, 2012. Used under license from Shutterstock.com; Page 4: © MindLeaders; Page 5: © MindLeaders; Page 6, 93: © MindLeaders ; Page 6, 117: © MindLeaders ; Page 7: © Dmitriy Shironosov, 2012. Used under license from Shutterstock.com.; Page 7, 46: © khz, 2012. Used under license from Shutterstock.com; Page 8: Image © Monkey Business Images;. Used under license from Shutterstock.com; Page 8, 53: © MindLeaders; Page 9: Courtesy of Thiago Felipe Festa; Page 9: © Andrey Pavlov, 2012. Used under license from Shutterstock.com; Page 9, 36, 45: © Peter G, 2012. Used under license from Shutterstock.com.

CHAPTER 2

Page 11: © Sebastian Kaulitzki, 2012. Used under license from Shutterstock.com; Page 12: Image Copyright John Wiley & Sons, Inc., Photographer J. Gerard Smith, 2011.; Page 15 (top right): Courtesy of Fran Flores; Page 15 (center): Courtesy of Christopher Hill; Page 15: © VIPDesign USA, 2012. Used under license from Shutterstock.com; Page 15, 57: © Luiz Rocha, 2012. Used under license from Shutterstock.com; Page 19: Courtesy of Lorenzo González; Page 20: © D. Kucharski & K. Kucharska, 2012. Used under license from Shutterstock.com; Page 21: © Czintos Odon, 2012. Used under license from Shutterstock.com; Page 23: © Discovod, 2012. Used under license from Shutterstock.com; Page 23: © Joerg Beuge, 2012. Used under license from Shutterstock.com.

CHAPTER 3

Page 27, 83: © Pinkcandy , 2012. Used under license from Shutterstock.com; Page 28: Courtesy of Gabriella Fabbri; Page 28: © Tamara Kulikova , 2012. Used under license from Shutterstock.com; Page 29: Courtesy of Nicolas Raymond: Page 31: © Kris Vandereycken, 2012. Used under license from Shutterstock.com; Page 31: Courtesy of Wootz; Page 34: Courtesy of Christina J.R. Hannah; Page 34: © S_E, 2012. Used under license from Shutterstock.com; Page 34: © Saharr, 2012. Used under license from Shutterstock.com; Page 34: © yellowrail, 2012. Used under license from Shutterstock.com.

CHAPTER 4

Page 37: © MindLeaders ; Page 38 (center): © doglikehorse, 2012. Used under license from Shutterstock.com; Page 38 (top left): Courtesy of Ginny Warner; Page 38 (bottom left): Courtesy of Spencer Ritenour via Park Slope Lens; Page 39 (top left): Courtesy of Simona Balint; Page 39 (top left): © Monkey Business Images, 2012. Used under license from Shutterstock.com; Page 39 (top right) : ©pinkcandy, 2012. Used under license from shutterstock.com.; Page 40: © Jan Danel, 2012. Used under license from Shut-terstock.com; Page 41: © MindLeaders ; Page 42: © michaeljung, 2012. Used under license from Shutterstock.com; Page 42: Kuzma, 2012. Used under license from Shutterstock.com; Page 44: All images © MindLeaders ; Page 45: © Muriel, 2012. Used under license from Shutterstock.com.

CHAPTER 5

Page 49: © Valentyn Volkov, 2012. Used under license from Shutterstock.com; Page 49: © Olga Lyubkina, 2012. Used under license from Shutterstock.com; Page 49: © kzww, 2012. Used under license from Shutterstock.com; Page 50: ©Raul Garcia/Getty Images, Inc.; Page 52: © Robert Pernell, 2012. Used under license from Shutterstock.com; Page 52: © HSNphotography, 2012. Used under license from Shutterstock.com; Page 53: © Karam Miri, 2012. Used under license from shutterstock.com; Page 54: © Martin Diebel/Getty Images, Inc.; Page 55: Jonathan Feinstein, 2012. Used under license from Shutterstock.com; Page 56: © photocritical/iStockphoto; Page 57: © donatas1205, 2012. Used under license from Shutterstock.com; Page 60: © Agita Leimane, 2012. Used under license from Photos.com ; Page 62: © MindLeaders; Page 64: © imagestalk, 2012. Used under license from Shutterstock.com; Page 64: © PhotoObjects.net, 2012. Used under license from Photos.com; Page 68: © James "BO" Insogna, 2012. Used under license from Shutterstock.com; Page 70: Courtesy of Marta Juez.

CHAPTER 6

Page 72: © Svetlana Yudina, 2012. Used under license from Shutterstock.com; Page 74 (bottom center): Courtesy of Elvis Santana; Page 74 (bottom right): Courtesy of Elvis Santana; Page 79: © MindLeaders. Page 95: © Evgeny Litinov, 2012. Used under license from Shutterstock.com.

CHAPTER 7

Page 85: © MindLeaders; Page 86: © MindLeaders; Page 88: © Fedor Kondratenko, 2012. Used under license from Shutterstock.com. Page 89: All photos © MindLeaders; Page 90: © MindLeaders; Page 91: © MindLeaders.

CHAPTER 8

Page 95: © Anson0618, 2012. Used under license from Shutterstock.com; Page 97: © MindLeaders; Page 98: Used under license from Photos.com; Page 94: © Anson0618 , 2012. Used under license from Shutterstock.com; Page 97: © MindLeaders; Page 98: © MindLeaders; Page 99: © MindLeaders; Page 100: All photos © Jupiterimages, 2012. Used under license from Photos.com; Page 101: ©Huntstock/Getty Images, Inc.; Page 102: © MindLeaders.

CHAPTER 9

Page 112: All photos © MindLeaders; Page 113: © MindLeaders; Page 114: All photos © MindLeaders; Page 116: © MindLeaders; Page 119: © MindLeaders.

APPENDIX C

Page 126: © MindLeaders.

APPENDIX D

Page 127: © MindLeaders.

FROZEN FOOD

CELL
nucleus (Eye

Liquid

Nutrition

THE value of Nutrients

Find out what Nutrition is   on WEDNSDAY

M-I NAMED Not SIRUS —

HIV n HEPATITIS A

~~BAT~~OUGH IM STILL A SUBMICROSCOPIC PARASITE

IN SHORT ~~ANYWAY~~ CALL ME VIRUS

FIRST NAME NORO LAST NAME VIRUS

WHEN ADDRESSING ME ~~ORO~~PROFESSIONALLY

CALL ME NOROVIRUS

~~ANYWAY~~

NEVER BEEN A MINOR

IM IN THE MAJOR League OF FOOD BORNE ILLNESS

AND THE ~~PRINCIPAL~~ YO DOCTOCTOR CALLS ME

VIRO GASTROENTERITIS

YES MR. MRS. ~~ADDRESS~~ AND MISS

~~LIKE GOD~~ IM NO RESPECT IN PERSONS OR PREJUDICE

BUT THE DIFFERENCE BETWEEN GOD AND ~~NOVIRUS~~NOROVIRUS

~~MY NOROVIRUS~~ ~~MEWAL POISON INFECTION IS~~

~~IMMM INFECTION NEW~~

GOD BRINGS PROTECTION I BRING VIRAL INFECTION

I COME BY SEA n LAND ~~FROM~~ INFECTING SHELLFISH

FROM DIRTY WATER OR HITCH HIKED ~~ON DIRTY HANDS~~

TO YOUR ENVIONMENT BY ~~DIRTY HANDS~~ SOME DIRTY HANDS

I EXIST IN FECES' DID THE TOILET PAPER MISS

DID YOU WASH YOUR HANDS FOR 20 SECONDS AFTER THAT

~~10~~ MINUTE SIT.

D-DAY - 2 DAYS LATER IVE GOT YOU STUNG

PEPTOBISMO CANT HELP, ~~YOU~~ GOT THE RUNS

WHATS THE RUNS DIARREH SON

NOROVIRUS ~~THE~~ THE ONE